HORSE CARE & RIDING
A THINKING APPROACH

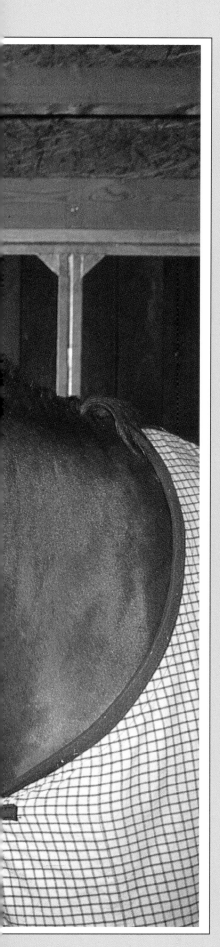

SUSAN McBANE

HORSE CARE & RIDING

A THINKING APPROACH

DAVID & CHARLES

This book is dedicated to the memory of
JACK FISHER
who died just before it went to press
I valued his friendship for many years. He was kind,
loyal, clever, extrovert and a very special person.

Books by Susan McBane
Keeping a Horse Outdoors (David & Charles)
Your First Horse: a guide to Buying and Owning
Keeping Horses: how to save Time and Money
Behaviour Problems in Horses (David & Charles)
Ponywise
The Horse and The Bit (editor)
The Horse in Winter
Effective Horse and Pony Management: a Failsafe System
    (David & Charles)
A Natural Approach to Horse Management
The Competition Horse: its Breeding, Production and
    Management (co-author)
Horse Facts (co-author)
Know Your Pony
Understanding Your Horse
Pony Problems: how to cope

Contributor to
The Octopus Complete Book of the Horse
The Pony Book
The New Book of the Horse

Most photographs by the author with additional
photography by Bob Langrish, with thanks to Janet Lorch
and the staff of the Muschamp Stud, especially Grania

Line illustration by Joy Claxton

British Library Cataloguing in Publication Data
McBane, Susan
  Horse care and riding
  1. Livestock: Care & training
  I. Title
636. 1083

ISBN 0-7153-9316-2 (Hardback)
ISBN 0-7153-0220-5 (Paperback)
First published in 1991
Reprinted 1993
First published in paperback 1994
Reprinted 1995

Typeset by ABM Typographics Ltd, Hull
and printed in Hong Kong by Wing King Tong Co. Ltd.
for David & Charles plc
Brunel House   Newton Abbot   Devon

EQUESTRIAN TITLES FROM DAVID & CHARLES
BEHAVIOUR PROBLEMS IN HORSES: *Susan McBane*
COMPETITION TRAINING
For Horse and Rider: *Monty Mortimer*
DAVID BROOME'S TRAINING MANUAL: *Marcy Pavord*
FITNESS FOR HORSE AND RIDER: *Jane Holderness-Roddam*
FROM FOAL TO FULL-GROWN: *Janet Lorch*
HORSE BREEDING: *Peter D. Rossdale*
HORSE CARE & RIDING
A Thinking Approach: *Susan McBane*
THE HORSE OWNER'S HANDBOOK: *Monty Mortimer*
THE HORSE RIDER'S HANDBOOK: *Monty Mortimer*
THE HORSE'S HEALTH FROM A TO Z
An Equine Veterinary Dictionary:
Peter D. Rossdale and Susan M. Wreford
THE ILLUSTRATED GUIDE TO HORSE TACK:
Susan McBane
KEEPING A HORSE OUTDOORS: *Susan McBane*
THE LESS-THAN-PERFECT RIDER
Overcoming Common Riding Problems:
Lesley Bayley and Caroline Davis
LUNGEING
The Horse and Rider: *Sheila Inderwick*
MARY THOMSON'S EVENTING YEAR
A Month-by-Month Plan for Training a Champion:
Debby Sly with Mary Thomson
A PHOTOGRAPHIC GUIDE TO STABLE MANAGEMENT:
Robert Oliver and Bob Langrish
PRACTICAL DRESSAGE: *Jane Kidd*
THE RIDING INSTRUCTOR'S HANDBOOK:
Monty Mortimer
ROBERT SMITH'S YOUNG SHOWJUMPER
Selecting · Training · Competing: *Rachel Lambert*
THE STABLE VETERINARY HANDBOOK: *Colin Vogel*
UNDERSTANDING HORSES
The Key to Success: *Garda Langley*

Other titles are described in more detail in the section on
further reading.

CONTENTS

Everything to do with horses, their riding and management, is obviously very practical. There are very many good riding schools and centres where people can learn to ride, and also learn about stable management, as horse care is called (though probably to a lesser extent unless they are on a formal course). However, such is the wide-ranging and complex nature of the subject that nothing but good can come from supplementing, or even preparing your time spent at a riding centre or with your own first horse or pony, with practical reading at a suitable level. This book is intended as a foundation for not-terribly-experienced people in the increasingly popular subject of horse and pony care and management. I believe that it will go a long way towards explaining many of the apparently confusing practices often witnessed by novices, and many of the pitfalls they may not even know exist. Further, because confidence stems from knowledge, I feel it will increase that of all those people who may find themselves groping in the dark, as it were, in their attempts to handle safely, effectively and in all fairness an animal which is large, heavy and strong with a will of its own and very specific physical functions and mental processes.

Unfortunately, there is a lot of traditional dogma put about in the horse world, both in practice, in lessons and in some books, which is based on flawed theories, lack of practical and scientific understanding of the horse and, sometimes, an unwillingness to consider the possibility that one might be wrong or that there might at least be another, perhaps better, way of doing something – better for both human *and* horse.

The material presented in this book is based on not only proven principles, the latest thinking and research and a sincere consideration for the horse, but also on a blend of the best of the old and the new. I hope very much that it will be referred to again and again as a base to which to return in times of doubt. So often when things go wrong it is because some basic fact or practice has not been thoroughly mastered.

The publishers of this book, David & Charles, are noted for their down to earth, practical books on a variety of subjects, and they have an extensive list of horse books to choose from for further reading. In order to include all the subjects which had to be covered the content of this particular book is very wide ranging, and for the most part this has meant giving no more than the basic, most important facts needed to provide a sound basis for the understanding of a topic. This does, however, mean that novice readers are not bogged down with information which is too advanced for them yet to be able to grasp or, in fact, needed at this stage.

The wisest and most experienced horsepeople know that you never stop learning about horses. They are very complex animals offering great aesthetic rewards to those who genuinely care about them, and who treat them kindly and correctly. I hope this book will act as a sound foundation for further rewarding study and enjoyment of them.

PREFACE

Whatever kind of horse or pony you want, rest assured it exists. It may not be easy to find, but the vast assortment of man-made breeds, 'mongrels' (placed in inverted commas because it is not a correct equestrian term) and types ensures that the horse for you is out there somewhere – it's simply a matter of finding it. This concise run-down of what is available will be a guide to the attributes, qualities and also disadvantages of what may seem to the novice to be a bewildering selection of equine merchandise.

BREEDS AND TYPES

Like all creatures, horses and ponies evolved according to the environment and climate in the region they happened to inhabit. As an animal with definite nomadic tendencies, however, the horse family (including zebras and asses) has spread to most areas of the world, though its 'birthplace' is generally taken to be North America, where the oldest fossils found to date were discovered.

From a general observation of horses and ponies at, say, a local show it is obvious that there are big horses, tiny ponies, hairy ones and sleek ones, and everything in between. To simplify the selection however, we can say that basically certain equine populations evolved largely in cold regions of the world, and others in hot regions, and that there are variations in between.

Britain's native mountain and moorland ponies and cobs are an excellent example of animals which have evolved in a cold-ish climate. We are classed as a temperate country (climate-wise!) but must remember that we do inhabit northern Europe and that Shetland is only 400 miles south of the Arctic Circle, although the North Atlantic Current or Gulf Stream does make our climate milder than may be expected for our latitudes.

Our natives, then, have thick, long coats in winter to protect them from cold winds and rain, and also from freezing temperatures, snow and ice. The Shetland pony and some other natives have a so-called 'double coat', when longish, coarse hairs intermingle with softer, shorter under-hairs; the top or longer hairs protect against the weather and drain away rain, while the softer undercoat helps conserve body heat. The thick mane and tail hair typical of these breeds also give protection. In summer they cast (moult) their coats and grow a shorter, sleeker summer coat – not only is protection not needed as much but the animals would, in fact, suffer from the heat if burdened with such a winter coat.

Other northern breeds work on the same principle, as do the large draught-type animals such as Shires and Percherons, and those other farm breeds which evolved in the bitterly cold steppes and tundras of middle and northern East Europe. Such breeds and types are termed 'cold blooded' – this does not mean that their body temperature is cooler than other animals, but that their temperament is normally rather phlegmatic and not so excitable as the so-called hot-blooded breeds.

These hot-bloods include the Arab, its descendant the Thoroughbred and, many would say, the little Caspian and the Barb. With the exception of the Thoroughbred, these breeds developed in climatically hot areas of the world, the earliest knowledge of them, as revealed both from fossils and in historical records, relating to ancient civilisations such as the Assyrians, the Persians and the Chaldeans. The Middle East, then, is believed to the the birthplace of these horses.

As well as usually having more sensitive and excitable natures than their less finely bred cousins, hot-bloods also grow much shorter, finer winter coats as historically they have not needed particularly heavy winter protection; however, effecient heat loss has been essential to them, therefore they have thinner skin, shorter hair and a more generous blood supply near the surface of the skin, all of which enable heat to radiate out of the body more easily.

The Thoroughbred, probably the most famous man-made breed in the world, is mainly descended from the Arab, though there are other oriental types in its makeup and also traces of native and naturalised, colder blooded British stock, for

A lovely type of Dutch warmblood of the sort becoming increasingly popular for competition work at all levels. Warmbloods normally have more placid temperaments than Thoroughbreds with elastic, long strides of the type particularly favoured now in dressage and show jumping. However, Thoroughbreds are still a favourite breed for eventing because of their boldness and speed.

Shetland Ponies and their crosses are popular as children's ponies. Dudley here is a Shetland cross with none of the 'Sheltie tricks' some owners complain of such as naughtiness and bad temper. His owners say they believe it is because he has always been treated with fair discipline like a horse, not like a cuddly toy.

it was created in England only two or three hundred years ago, initially in Yorkshire. Because of its mixed ancestry there is considerable variety within the breed, some individuals growing heavier winter coats and looking decidedly more 'common' than others. The Thoroughbred, and particularly the Arab, have both been used to improve or 'breed up' just about every riding and light carriage breed in the world, and many ponies, cobs and heavier horse breeds, too.

In competition riding circles the continental European warmblood breeds are currently much in fashion; these, certainly, have much Thoroughbred (and therefore Arab) blood in them, which has been crossed with the native horses and ponies of the different individual countries. Man-made warmblood breeds such as the Hanoverian, the Trakehner, the Westphalian, the Dutch warmblood, the Danish warmblood, the Holsteiner and so on, are examples of what we term warmbloods.

British native ponies comprise the Shetland and the two types of Highland pony, the Fell from Cumbria, the Dales from Yorkshire, the two Welsh types of pony and the famous Welsh Cob with, in between, the Welsh Pony of Cob Type, then the New Forest. There is also a now-established breed called the Lundy Pony bred from the New Forest, and a rather super British native horse from Yorkshire, the Cleveland Bay, considered ideal for carriage driving. In the extreme south of the country are the Dartmoor and Exmoor ponies, the latter being the nearest to the pure primitive type of pony which first evolved in our climate, the others having had admixtures of other types – including the Arab – over the centuries.

Ireland developed the renowned Connemara pony and also the Irish Draught horse, the latter being particularly well known for crossing with the Thoroughbred in various proportions to produce superb hunters and competition horses, particularly jumpers.

CONFORMATION

Conformation (make and shape) is a tricky subject about which whole books have been written. Your adviser will check the conformation of any horse you are considering buying, but basically you should look for a horse who seems to 'fit' within himself, who has good overall balance and symmetry and is neither weedy nor coarse and 'cloddy'. Developing 'an eye for a horse', as acquiring good judgement of horses is called, is not that difficult: start by attending good horse shows, such as County level and above, and study the winners of the showing classes till you have imprinted on your mind's eye what a well made, balanced individual looks like. Also study different types such as hacks, hunters and cobs, and also different breeds such as Arabs, Welsh Cobs and Irish Draughts so you become almost instinctively able to recognise them and their crosses.

The horse pictured here does have good balance in his conformation. He is well built (called 'having substance') but not heavy or coarse. He is not elegant or beautiful but handsome and 'workmanlike' with a kind expression in his eye and on his face which denotes he probably has a good temperament.

Some points to note are that the croup should not be higher than the withers, that the point of the hock should be no higher than the chestnut inside the foreleg on the same side, that the horse has an open elbow (you should be able to fit about three fingers widthways between the elbow and the ribcage) and that the horse is 'well ribbed up', meaning you should just be able to fit the width of your hand between his last rib (you'll have to feel for it) and the point of his hip. From the back, his gaskins should be the same width as his hips. The shoulders should be only slightly narrower than the hips.

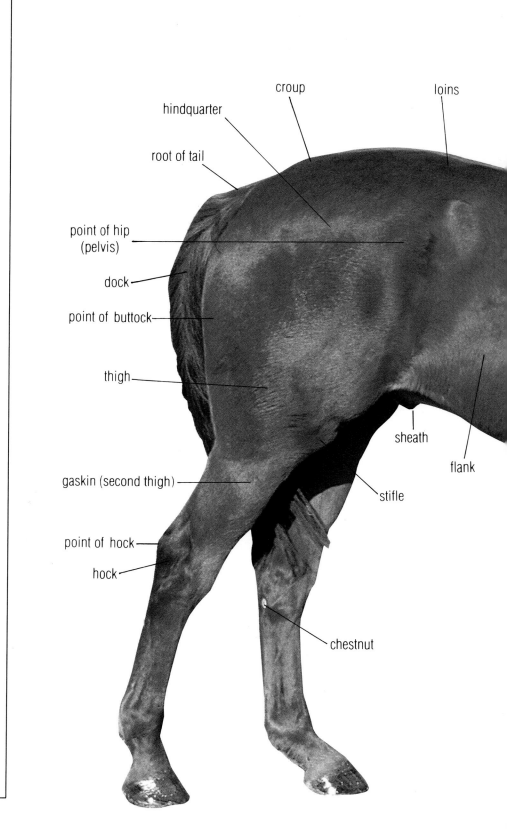

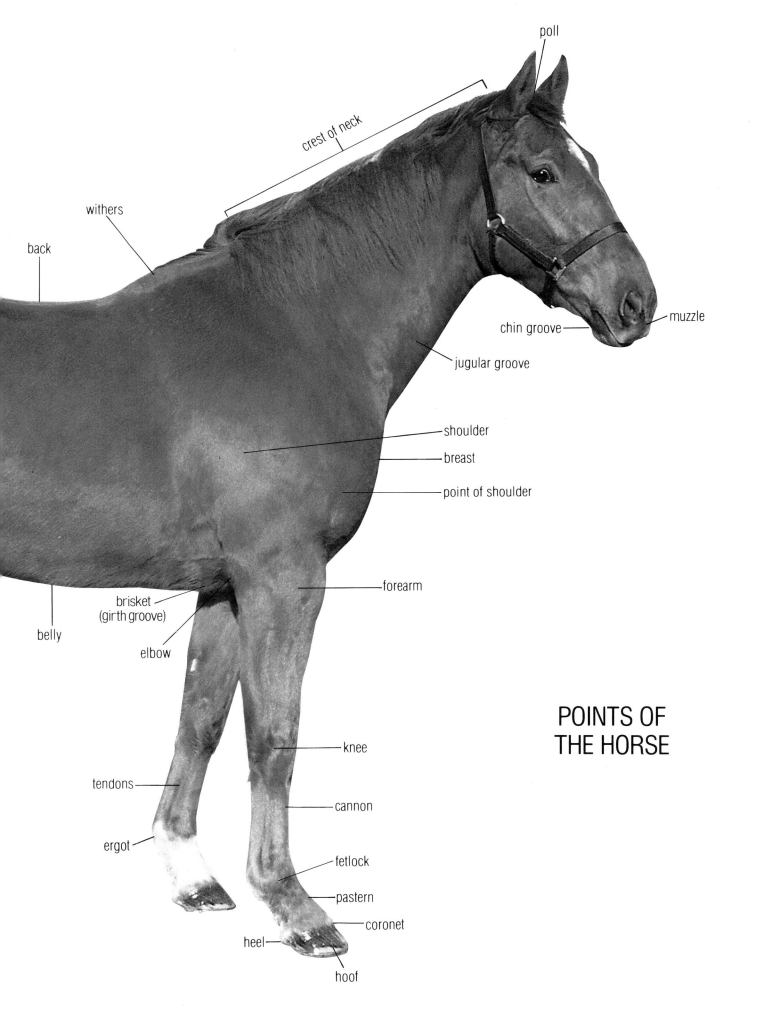

poll

crest of neck

withers

back

chin groove

muzzle

jugular groove

shoulder

breast

point of shoulder

forearm

brisket
(girth groove)

belly

elbow

POINTS OF
THE HORSE

knee

tendons

cannon

ergot

fetlock

pastern

coronet

heel

hoof

A piebald (black and white) cob-type pony, stocky in build and suitable for older children or adults, able to carry more weight than the pony shown opposite.

HORSES DOWN UNDER

Australia and New Zealand have no native horses or ponies, and neither has America – despite the wide range of breeds and types existing there now, they are all man-made from imported stock. After the horse's ancestors migrated, millions of years ago, across the Bering land bridge (now the Bering Straits between Alaska and Russia) and, some palaeozoologists believe, possibly across the newly forming, shallow Atlantic ocean, the horse family died out in America for unknown reasons and was reintroduced there only when the Spanish conquistadores landed. Australasia broke away from the remaining land mass many millions of years before any horse family ancestors migrated there; it received its first sight of horses when white races began to settle there. The most famous breed now is the Australian Waler.

THE ELEGANT HACK

The hack is a very refined, elegant, good-looking-to-beautiful animal which is supposed to have faultless manners (not quite opening doors for you) and be schooled to virtually advanced dressage standard – in short, an effortless pleasure to ride, guaranteed not to play up and, at least in the old days, the sort of horse you'd be proud to be seen riding in the park. Today we do not see quite the refinement and elegance we used to, and show hacks are not so often ridden with one hand (by gentlemen, at least, leaving the other free to raise his hat to a lady). A hack used to be defined as any good riding horse not capable of following hounds (perhaps it became crazy at the sight of hounds, or would not jump, or simply had neither the gallop nor stamina to perform across country). Riding schools and livery stables in the old days frequently advertised 'Hacks for Hire', something never seen nowadays; today a hack usually means a show hack, not a general riding horse.

America has very many breeds, all derived from introduced horses and ponies, including the ubiquitous Thoroughbred and Arab. There are Quarter Horses, Saddlebreds, Standardbreds, Walking Horses, Morgans, the Pony of the Americas, the American Shetland and many, many more, including draught breeds for farm work and logging. The smart Appaloosa breed was developed in the Palouse River region, initially by the Nez Percé Indians.

Here in Britain, horses which are considered *types* rather than breeds are the riding cob, the hunter and the hack. Such animals often have no pedigree or registration papers (excluding the Welsh Cob) and often the breeding is unknown or known only a generation or two back.

A cob is a medium-sized horse, no taller than 15.2 hands high (a hand being four inches), of stocky build with shortish legs, strong quarters and an attractive head and neck which should be neither fine nor coarse. Generally it has a patient, willing temperament and it makes an ideal family first horse, able to carry adults and older children alike or go in harness to perhaps a smart trap, or whatever you wish. The old fashioned 'ride and drive' animal was very often a cob, and still is.

A hunter is obstensibly any animal you can ride to hounds, from a Shetland to a Thoroughbred. However, the British hunter is a well defined type divided into three categories particularly in showing circles; heavyweight, middleweight and lightweight, depending on the weight the horse is felt able to carry for a long day with hounds, possibly five hours or so. The best way to train your eye, in a nutshell, is to visit as many shows at county, regional and national level as you can, and study the type of horse being exhibited in the various class designations. You'll soon develop a mental picture. Also, look in horse magazines such as *Horse and Hound* which feature show results, and study the pictures of show winners to train your eye to recognise the best of each type.

The riding horse is another type of horse which, again, in theory covers almost anything you can sit on but which in practice is becoming a fairly well-defined type. It does not fall into a clear cob, hunter or hack category but is nevertheless a pleasant, good-looking animal to ride. In fact, many riding horse classes are filled with middleweight, and sometimes even lightweight hunters although some of the exhibits would obviously be 'off type' for a high-class hunter class.

The horse world does not have an equivalent word for mongrel, although it is obvious that the majority of animals bought and sold for general riding are, in practice, just that – 'Heinz 57' types of unprovable breeding but doing a good job of work. Competition horse breeding here is catching up with continental Europe, where 'papers' are everything and potential competition horses go through strict inspection shows (called 'grading') before being granted registration papers, which acknowledge they have reached a certain standard, and recognise them as 'fully paid up members' of their breed. Animals not coming up to scratch are not granted papers and are not accepted as approved breed representatives. Many British competition horses have huge gaps in their pedigrees – but handsome is as handsome does in many cases, and in this country you can get a good competition horse with no papers at all.

WHERE ARE YOU GOING TO KEEP HIM?

If you have reached the stage where you are seriously considering buying a horse or pony, and have perhaps been attending a good riding school for some time to attain a reasonable level of expertise, you will already know what kind of accommodation equines need. It is not possible to keep one in your converted garden shed, your garage, an old hen house or just on your back garden. Horses are big, strong animals with a natural tendency to and craving for space and movement. People do keep horses and ponies in all sorts of unsuitable conditions, of course, but at risk to their own safety and the horse's wellbeing.

A fairly ordinary sort of 'mongrel' family pony of substantial build but not heavy or coarse and with an alert but placid temperament.

BUYING HORSES AND PONIES

Where are you going to keep him?
Finding your horse

KEEPING YOUR HORSE HAPPY

There is one very important aspect of keeping a horse. Nearly all horses need the company of their own kind if they are to be happy, contented, feel secure and, therefore, thrive mentally and physically. If you are keeping your horse at home you will almost certainly have to provide another equine for company, either by getting a small pony or by letting off another stable to someone else – though then you may have problems with the two of them becoming impossibly attached to each other! Ideally, horses should be kept in numbers of three or four upwards, although this is not essential.

If you plump for a riding school, livery stable or other rented accommodation you should at least be assured of company for your horse.

'American barn'-type stabling. Popular in areas of inclement weather conditions, this type (with screening) enables a horse to have safe contact with another

For most privately kept horses both stabling and grazing are needed. True, many animals live with only one or the other, but this is far from ideal. Even if a horse lives out most of the year there are times when a stable is needed, such as when he is sick or injured; it's also much easier to prepare him for a show, for example, or to confine him for a vet's inspection or for clipping or shoeing if he can be at least temporarily stabled. And most horses do like to come inside at least at night in winter and during the day in summer, particularly if there are no adequate shelter facilities in their field.

Most privately owned horses are deprived of sufficient exercise. Their owners attend school, college, they work or they have family and other commitments which keep them from giving their horses enough exercise – many are lucky if they get an hour a day. And the two hours normally recommended for a healthy, fairly fit horse still does not come anywhere near the many hours of gentle exercise taken by horses free to wander at will; and in any case, enjoyable though work-type exercise may be (but often is not), there is nothing quite like liberty for the horse's mental contentment and health.

Ideally, to satisfy their inborn need for space and movement, horses should have large fields of about ten acres upwards – but small paddocks of a quarter-acre are better than nothing, or even just a well-drained outdoor manège of the usual 20 by 40 metres where they can be turned out to stretch their legs when paddocks are waterlogged. Comparatively modest facilities like this, combined with ridden or driven exercise, make life happier and healthier for the horse and easier for those responsible for his care.

In addition to stabling and turn-out facilities, you'll need somewhere to store feed and bedding, and items such as hay or hayage and straw or shavings can take up a lot of space if bought in larger, economically sized loads; on the whole, and within reason, the more you buy the less it costs per unit (per ton of hay, per bale of straw). It is certainly possible to manage with a covered storage area only the size of a stable or garage, but you will find yourself paying more for bulky items like hay.

Tack, clothing, grooming kit and veterinary supplies also need dry storage, if possible at room temperature. Leather items in particular soon deteriorate and go mouldy if kept in cool, damp places. You can always use part of a spare room at home, but most stable complexes include a special tack room which, for only one or two horses, need be no bigger than a stable or even less.

Bearing in mind this brief run-down of the basic accommodation needed to keep a horse, you'll have to decide if you *can* provide it at home if that is where you want to keep yours. Otherwise it will give you a fair idea of what to look for in any livery stables you may visit, or in rented accommodation you may inspect with a view to boarding your horse out.

FINDING YOUR HORSE

One of the most satisfactory ways for a comparative novice to find a suitable horse is to go to a reputable dealer; your horse consultant should be able to put you in touch with one. Unfortunately dealers don't have the best of reputations, but a good one is a different matter. For a start, they should have several suitable animals on the premises for you to try, something you won't get anywhere else. They often allow trials at home (insured at your expense) and will take back an unsuitable animal even after payment (sometimes less a percentage for their time and goodwill) and change it for another or, depending on the firm, refund a fair proportion of your money. If you already have a horse they'll often part-exchange it for another or even do a straight swap, depending on the animals involved.

They have a good reputation to maintain, and although they are business people working for profit (like anyone else) they have to abide by consumer laws and operate honestly and fairly if they want to stay in business. Word travels fast and far in the horse world and a dealer who constantly sells duds or tries to swindle his customers won't last long *or* be successful, unless the customer is equally wily!

Dealers will be known locally or regionally. They may be recommended by word of mouth by your consultant, by the riding school you have been attending or by other horse owners, or you may see them advertising in the equestrian press, and you will also find a listing of them in the *British Equestrian Directory* available from Equestrian Management Consultants Ltd, Wothersome Grange, Bramham, Wetherby, West Yorkshire, LS23 6LY. This directory is invaluable as a fact-finder and contact point for just about everything and everyone you'll need in the horse world. It is very comprehensive, and is an excellent starting point when you want information.

Sales

This is an extremely risky way of buying a horse unless you are very knowledgeable – even if your consultant qualifies, he or she may still prefer to give this source of supply a miss. Even if the horse you like is warranted you cannot normally try it beforehand, except at some specialist sales and the better quality non-specialist ones. Good sales permit you to return some categories of animal within a stated time should they prove to be not as represented nor what you might reasonably have expected, but this point must be very carefully cleared up before you start waving your catalogue about or scratching your nose, all of which may be taken as a bid!

A reputable sale is quite safe to buy from, but your consultant should help you check out its credentials before taking the plunge. Sales are advertised regularly in regional and equestrian journals.

Riding Schools

If you've been attending a particular school you may already be in love with your regular mount who, in turn, may be ready for retirement into private life or just

PERFECT MISERY

Adequate shelter facilities are so important yet so often overlooked or disregarded. Forcing horses to endure conditions like this is extremely poor horsemastership tantamount, in the author's opinion, to cruelty. There is no shelter, the ground is waterlogged and despite possibly a winter coat, the horse has no real defence against the chilling effects of wind, rain and cold, with not even company for moral support. Horses can and do die from conditions like this after weeks and months of intense suffering from exposure. No amount of food can compensate for the excessively high demands on the horse's inner resources enforced by enduring this sort of treatment.

This is the other side of the coin – a horse driven out of his mind by the attacks and irritations of flies. Again, with no shelter the horse is driven to galloping away from them, but the second he stops they attack again. Rock hard ground breaks his feet and jars his legs to the point of lameness which takes away his only defence. Galloping produces sweating which attracts even more flies and, in any case, cannot, obviously, be kept up for long. Shelter is just as imperative in summer as in winter, from both flies and sun.

THE RIGHT TYPE FOR YOU

Much depends on the job you want your horse or pony to do. If you are a novice wanting a quiet, reliable, well-mannered horse to bring *you* on, you'll pay a lot less than if you are fairly competent and perhaps want to start competing. The 'starter' horse's qualities are just as vital for you as are scope and talent for the competitor, but they don't fetch as much in the marketplace.

Size and weight-carrying ability are really more important than breed or specific type unless you particularly want a horse of a certain category. There is an old guide which says that a horse is capable of carrying happily and working reasonably with a rider 12-15 per cent of his own weight. This would mean that a typical riding club type of horse — say, three-quarter Thoroughbred, and weighing 1,000lb — would pair well with a rider of about 9 or 10 stones in weight, which is about right in practice.

As for height, this really depends on what you are comfortable on. Many of those who advocate our native ponies claim that these can carry a man (meaning a farmer) all day — and they do seem able to, if you don't mind your feet nearly touching the ground. As a rough guide, when you are mounted on your prospective new horse, if the soles of your boots are about level with its sternum (breastbone) it is about the right height for you.

With both these factors, however, much depends on the build and constitution of the individual horse or pony: obviously a finely bred animal of 15 hands will not be up to the same weight as a 15-hand cob — although Arab en-thusiasts (of which I am one) would deny this, as the breed is well known for its strength and stamina and weight-carrying abi ity.

Temperament, too, is important. Ponies and cobs are usually more placid and tolerant than Thoroughbreds, Arabs and their cross, the Anglo-Arab, or animals with much of this breeding in them, but this is not always sc — I have known some very fiery Welsh Cobs for instance, and some really quiet and amenable Arabs. If you are of a nervous disposition, go for a placid type; on the other hand, if you are calm and not prone to panic or are even rather phlegmatic in temperament, you could perhaps do with something to wake you up a bit!

If you have shortish legs, don't go for a sturdy, barrel-like mount as you'll never be able to fit your legs round it and sit comfortably. Conversely, if you are tall and long-legged, a narrowly built horse will give you the feeling that you have nothing underneath you — not a feeling to inspire confidence.

As for sex, this book is being written for novices and I feel geldings (castrated males) are generally best for them. Stallions (uncastrated males) need expert handling, and although some do show a superb temperament in knowledgeable hands, in those of a novice they can become dangerous and unmanageable. Mares can also be difficult — they come into season every three weeks or so during the spring and summer and can be very trying then. It's all in the hormones, of course, and in general geldings have fewer problems in this direction than other animals.

If you have reached the stage where you want to go on a little and maybe start competing, you will need a horse with more ability and scope. Few things are more frightening than landing yourself with a horse which is too much for you — on the other hand there is nothing more frustrating when competing than a horse which cannot (or will not) keep up with your aspirations.

Most important of all, and whatever anyone else says, you must like the horse yourself. You're going to spend a lot of time with him or her, and may be looking after him partially or entirely yourself. He should be your friend, and how can you be friends with someone you don't like? Like meeting people, dogs or other animals, choosing a house or buying a car, there's a good chance that you'll *know* when you click with a certain horse. You'll see a certain look in his eye, an air about him, or just experience a good feeling — and that will be the one.

But on this note, a word of warning: unless you don't mind throwing good money after bad, never buy a crock just because you like him and feel sorry for him, because you'll need a bottomless bank balance to keep him in veterinary visits. People *do* do this sort of thing, and it's fine to be humane and charitable if you can afford it and don't really mind the horse not being able to work much of the time. However, if you do expect your horse to work for his keep, and money is a consideration (as it is for most of us), be very careful about the health of the horse you buy, and banish sentiment.

feel like a change from the rather humdrum, slightly anonymous life of a riding school. This might work well for you both, but be aware that the horse may possibly refuse to work alone, as at very many riding school animals never have to. Be sure on this point by trying him alone, preferably on a hack, in traffic, or by having him for a week at wherever you intend to keep him. If you intend to carry on keeping him at livery at the school, at least make sure he'll go out alone for you when you feel like a hack – and still get him vetted before purchase, even though you know him well, as you just do not know what problems might be simmering underneath the surface.

Private sellers

Private sellers are not bound by the same consumer laws as professionals such as dealers or riding school owners. For example, they can represent a horse to be safe in traffic as far as they are aware, yet when you get it home it turns out to be terrified of tractors. When tackled, the former owners can legitimately say that they were not aware of the animal's fear of tractors because it never displayed such fear in their ownership (because it never encountered tractors). You don't have such problems when buying from a professional, not only because he or she has a reputation to maintain, which a private individual does not, but because he is bound by consumer laws.

Even getting a private seller to warrant a horse does not entirely remove the problem, as that little qualifying phrase 'to the best of my knowledge and belief' (or suchlike) can be inserted, and it can take a long and expensive court battle to try and prove otherwise, often unsuccessfully.

Many private individuals are, of course, perfectly honest and have genuine animals to sell for genuine reasons; however, many do no come into this category, and when buying privately your horse consultant can be an absolute boon.

Equestrian press

Some horse magazines, particularly weeklies, advertise horses for sale. They may be national publications or regional ones, and some farming journals do, too – *The Farmer's Guardian*, for instance, has a thriving horse section with articles as well as advertisements. The trick here is to read between the lines and think what is *not* said about an animal. For example, if a horse is described as 'sound in wind and limb' it could be blind and have a weak heart. 'Easy to box, shoe and clip' sounds fine, but what about catching up from the field? Circle any likely sounding animals and show them to your consultant; then you, or she, can make an initial 'phone call to ask a few more questions, and perhaps arrange a visit.

TRIALS AND INSPECTIONS

When you have made a short-list of animals likely to be suitable you must of course go and try them out. People often advise trying at least six, and not to fall for the first one you see; but why not, if it's just what you want? Visit the best sounding one first, then if disappointed, work your way down your list. However, there's no point wasting time and money looking at others if you've seen just what you want or very nearly, and if you delay you may miss your first choice. If this is your first horse, or one which represents a step up and of which you have great hopes, you may be quite nervous and excited; nonetheless you must try to keep calm and objective – difficult!

When you are first shown the horse, whether in his field or in the stable, the first thing to note is his general demeanour: is he pleased to meet you, or suspicious, withdrawm or even aggressive? If any of the last three are apparent, stop right there. As a novice, you don't want any temperament problems, no matter what reason you may be given. If they are possible to sort out, fair enough, but that's not your task – you want a friendly, co-operative character.

Part 2 covers basic conformation (make and shape) and action (how the horse moves as an individual), so are not covered here; these topics normally take up the first part of the inspection.

While the horse is being prepared for inspection, having his rugs removed if worn, his headcollar or bridle put on and so on, note how he reacts to these slightly uncomfortable processes, whether or not he snatches and bites at his groom or owner or whether he puts up with everything with good grace. It must be said in the horse's favour that if he *does* seem irritable, note first how the person is handling him; if he or she is being rough and inconsiderate, or rushing the horse, it's no wonder he's a bit upset.

While this is going on, look quickly at the horse's box and notice whether or not there are scuff marks on the lower walls, signs of boards or bricks having been kicked in, or kick damage to the door. Note also whether ledges appear to have been chewed – and this includes the top of the door, even if it is covered with metal. Look also at the outside of the door – there may be a semi-circular scrape mark where the horse has been running his teeth along it out of boredom. All these things may indicate a restless temperament, although they can also be induced by over-confinement – the horse being kept stabled too long with insufficient exercise and liberty. Point them out quietly to your adviser (who will probably have spotted them anyway), then enquiries can be made as to whether this is the horse's normal box – perhaps some other horse is responsible. You may not get the truth,

BUYING HORSES AND PONIES
The right type for you
Trials and inspections

AGENCIES

Agencies, too, advertise in the equestrian press. Their purpose is to put sellers and buyers in contact with each other; normally they take a commission from the seller, but some also operate a search service which may or may not involve the payment of a retainer from the buyer. They are not dealers and have no selection of horses for you to try, but since they have their ear to the ground, they will often come up with the goods for you. They usually get their clients on both sides to sign a simple agreement which ultimately means you have no claim on them if the horse is not exactly what you wanted or if something is wrong or goes amiss with the deal; and this is fair enough when dealing with the vagaries of human nature and something as sensitive, unpredictable and expensive as a horse which does not even belong to them.

BREED SOCIETIES

Societies dealing with a specific breed are not usually a good source of supply for novices, as they can often only put you in touch with breeders who, in turn, normally have only youngstock for sale. You need a mature animal which can teach *you*, not one needing the schooling which you are not yet competent to give. However, some breed societies do have an agency section which has a list of sellers with more mature pure and part-breds of their particular breed, so if you want a special breed contact its society, which will be listed in the *British Equestrian Directory* already mentioned.

BE FAIR

Normally sellers go to a lot of trouble to prepare a horse for inspection, so it is only fair to ring and cancel your appointment if you cannot make it for some reason. If you get a reputation as a time-waster you'll find it hard to get anyone to show you a good horse. Try to be punctual. It is often advised that you should arrive early so you can see if the horse is being warmed up (or worn down!) before his trial, to loosen up any stiffness or wear off lameness, or whatever. However, any problems should show themselves during the vet's inspection and may well be spotted by your adviser anyway. And if you do have the horse on trial it shouldn't be long before you discover any idiosyncracies it may have.

At the inspection and trial you will be guided by your adviser, but it *is* going to be your horse so don't be afraid to point things out or ask questions. Also, if you decide early on that this horse is not what you want, say so without delay, both to save the horse's energy and the seller's time.

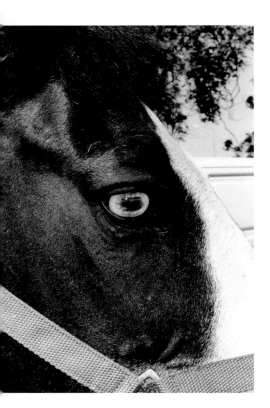

but after the trial, try to get a few minutes aside to watch the horse in his box and see if he does seem to have a behavioural problem.

When the horse is brought out and stripped (minus rugs) for your inspection, your adviser will probably step back to get an overall picture of him from a distance; she or he will consider both sides, to check whether the horse is a physically balanced individual, able to move freely, and will notice if he responds instantly and willingly when asked to come out and move around, rather than nervously and timidly, also whether he is interested in what is happening to him, if he is defensive or nervous.

When she or he has seen the horse from both sides, front and back, she will ask the leader to walk him away and trot him back. She will stand immediately behind him and check that each leg moves straight in the other's flight path and doesn't swing to the sides, and the same when the horse trots back. From the side, she will look to see whether the movement is free and flowing, that the horse reaches well under his belly with his hind feet and out under his head with his forefeet, or whether he seems to go in a stilted, 'proppy' way.

This trotting up should take place on a level, hard surface such as a concrete path or a hard road, because this will show up any faulty action, soreness or slight lameness. To be really critical, the horse should be lunged in a circle on a hard surface at a working trot, as this will reveal even slight lameness or action faults. The detection of lameness *per se* is the vet's job, but as an experienced horsemaster your adviser will know at once if the horse is not going right and will point it out; in which case no further examination is worthwhile, certainly not under saddle.

Your adviser will next examine the horse closely; she will run her hands down his legs and along his back to check for any adverse reaction, and for any warm spots or old lumps and bumps which could indicate future weakness; she will pick up his feet to see that they are healthy and strong, and will look at the teeth to get an idea of his age and for uneven wear of the front teeth which could mean that the horse is an inveterate wood chewer or crib-biter (which can, of course be caused by bad management and faulty feeding). Crib-biting is a vice which once confirmed is incurable, at the time of writing, although research is going on to try and find a cure. The horse grasps a ledge, or even his own knee, and uses it to arch his neck and form a vacuum in his throat; then by the action of his throat muscles, he releases the vacuum, gulping air down into his stomach. This condition can be caused by the distress of inactivity and over-confinement, or by faulty feeding and resultant indigestion. Some experts maintain that the air swallowed may cause further indigestion, but opinions differ on this.

Your adviser will also check the horse for general health: bright eyes, sleek coat and alert reactions, interest in his surroundings, friendliness and general demeanour.

Next the horse will be ridden to show his paces. His first rider should be the owner or groom so you can see how he reacts to familiar people. Does he move on to the schooling area willingly, and appear to go happily and effortlessly for his rider? – or does he 'back-pedal', thrash his tail about, refuse jumps, refuse to strike off into canter on a particular leg or seem to pull? Some misbehaviour could be the result of bad riding, which your adviser would identify.

Next, I would get your adviser to ride him at all paces and at whatever obstacles the horse is said to be able to perform over. General school movements such as lengthening and shortening of stride, leg-yielding, transitions to and from all paces, figures of eight, circles and so on should be tried so she can give you a quick report of the horse's ride – and then it's your turn. If you are getting a small horse or a pony and your adviser is too large to ride it, obviously this phase of the inspection is omitted.

Allow yourself a minute to sit in the saddle and get the feel of the horse at the halt. Do your legs fit round him nicely and are you comfortable? Does he feel too big or small for you, too wide or too narrow? Do you feel you have 'plenty in front

of you', or do you feel as though you are too close to his ears, as they say? Only you can tell how *you* find a horse. When you've sorted out your reins and stirrups, move off at a walk on a longish rein. Relax and ride naturally as best you can, remembering that you and the horse are total strangers. Try him on both reins at all paces, with changes of direction at all paces, and do whatever school movements and jumps you wish. Maybe the horse can do more than you can, which is fair enough as there'll be that much extra for your instructor to bring out, and the horse can help to teach you. Later on, when you're more experienced, you may wish to get a greener (less experienced) horse and school him – under expert supervision – but as a novice, a well schooled equine teacher is ideal for you.

If you like the ride the horse gives you and you feel you like him as a personality, there's no need to play cagey and say you'll think about it. Tell your adviser and the vendor how you feel, though *don't*, at this stage, say you'll have him. Your adviser or you yourself may ask further questions about the horse, to find out if there are any little quirks you ought to know about, such as (bearing in mind that no horse is perfect!) a dislike of cats, a tendency to roll in water or to be possessive

When running up a horse in hand, it is essential that it is done on a loose rein whether it is for inspection for purchase or a lameness check. A tight leadrope can easily disguise the action to the extent that the true gaits cannot be discerned. Given complete freedom of head and neck the horse will move naturally – and reveal all!

LEFT:
This pony has what is called a 'wall eye', in other words its pigment is blue instead of the normal brown. This does *not* indicate faulty sight but many people do not like its appearance, particularly in the show ring. Normally, only one eye is blue although some white horses have two blue eyes. True albino animals have pink eyes.

GETTING EXPERT HELP

Riding is a risky sport. If certain precautions are not taken it can be dangerous: insurance companies rate it as dangerous as motor racing and mountaineering, and some consider it more so. When you are actually riding, protective clothing such as a hard hat and a back protector are the sensible things to wear; but it is also wise to protect yourself before you even buy a horse by getting expert help and advice on various aspects of the procedure. You will have to pay for such professional advice but it is not an ongoing expense, and as far as the suitability of the horse is concerned it could save your life. It could also save you a great deal of money and disappointment.

Equestrian consultant

The two main equestrian organisations in the UK which offer various sorts of help to horse enthusiasts of all levels are the British Horse Society, at the British Equestrian Centre, Stoneleigh, Kenilworth, Warwickshire, CV8 2LR; and the Association of British Riding Schools, Old Brewery Yard, Penzance, Cornwall, TR18 2SL. They both run examinations and career structures which produce qualified people at all levels. An equestrian or horse consultant can probably help you find a horse, will inspect it for ability and suitability, assess your capabilities, and advise whether or not the two of you are well matched; he or she will also attend when you yourself go to try the horse. He will know what questions to ask and just how to try out the horse, will discuss such things as home trials, warranties and registration papers if any, and will also advise you on transporting the horse home and even on a good place to keep the horse.

The qualifications of consultants vary, but you should look for people with one or more of the following: the Association of British Riding Schools (ABRS) awards a Fellowship to very highly knowledgeable people, and so does the British Horse Society – the letters FABRS or FBHS after someone's name should therefore put you in very safe hands, as will the letters RSPD (the Riding School Principal's Diploma of the ABRS) and BHSI (the full British Horse Society Instructor certificate). There are also less highly qualified people who are still perfectly competent to advise you: they will use the letters ABRSGDip (the full Groom's Diploma of the ABRS) and/or BHSII (the British Horse Society Intermediate Instructor certificate).

There are experts with no qualifications and professionals not so highly qualified who may also do a good job: however, you should not take chances in an area in which you yourself are not very knowledgeable, so by keeping to the qualifications given above I feel you will be playing safe.

Veterinary surgeon

Your equestrian consultant is not (normally) a veterinary surgeon as well, and although he or she can certainly tell whether or not a horse is in good health, the task of giving the horse a full medical examination is the vet's – he will also give an opinion as to whether or not any disorders it may have are likely to affect your use of it. He or she will examine the horse thoroughly and give you a detailed and thoroughly professional opinion on the horse's state of health.

No horse is perfect and much depends on what job you plan for your mount. For example one sound (healthy enough) for active hacking and showing may not be sound enough for jumping or long distance riding, so you must be quite clear in your own mind what you really intend to do with the horse, and tell the vet accordingly – if you said you wanted a hack, he or she cannot be blamed if the horse later becomes severely lame because you have been competing regularly in jumping events.

Veterinary surgeons all have the same basic training but some subsequently specialise in one type of animal – small animals (dogs, cats, rabbits etc), or agricultural animals (sheep, cattle and so on), or mainly or even solely in horses – but you cannot tell this from their qualifications. All vets have the letters MRCVS (Member of the Royal College of Veterinary Surgeons) after their name, but they will often have other qualifications too, depending on the exact degree offered by the university they attended, such as Vet MB, B Vet Med or BVSc. Some will also have the Fellowship of the Royal College of Veterinary Surgeons (FRCVS) awarded to them.

To find an equine veterinary surgeon practising in the same area as the horse you are considering buying, write to the British Equine Veterinary Association, Hartham Park, Corsham, Wiltshire, SN13 0QB. BEVA is an association for those vets particularly interested in working with horses, so although no special equine qualification exists, membership of BEVA indicates considerable interest and expertise in horses, which is just what you need. Although the association will not actually recommend someone, it will be able to tell you which of its members practises where.

As the vet, like any other expert, will charge for his or her services, for travelling expenses and also sometimes for travelling time, it makes sense to pick one living near your prospective horse. However, you cannot use the vet used by the horse's owner as this will obviously involve a conflict of interest. When you tell the vet where the horse is, he will tell you if it belongs to one of his clients; if he cannot act for you, you'll have to contact someone else.

Many people try to economise on consultants, but the veterinary surgeon is one I particularly feel is absolutely essential: you should never buy an animal which has not been 'passed by the vet', as the term goes, for the job you have in mind.

Solicitor

The sale of horses is covered by the Trade Descriptions Act just like any other item for sale. Horses, like washing machines and anything else, must be of merchantable quality – in other words they must be fit for the purpose for which they are purchased. You *are* taking a chance with a living creature that something will go wrong with it in the future (and your vet's examination cannot, of course, confirm that nothing will) but you do have a certain amount of security if you buy a horse with a warranty, a guarantee in other words. The best person to advise you on this, apart from your consultant, is a solicitor.

A solicitor will not attend the horse but will give advice on the wording of a warranty or any other legal matter on which you are uncertain. Horses sold at sales usually come with a warranty, but privately sold horses often do not. If you ask for one and are refused, be suspicious, although this is really an area for your horse consultant to sort out. A warranty guarantees the horse's suitablitity for the purpose for which it is being bought. For instance, a horse warranted 'quiet to ride' is suitable for a reasonably experienced rider and should go safely in company or alone. You really want a horse which is quiet to ride and warranted safe in *all* traffic (not just with cars), which is sound in wind, eyes, heart and action and is free from disease and vices, also that has not been operated on for unsoundness of any kind unless this is declared by the seller.

The whole area of warranties and conditions is quite complex and should be thoroughly discussed, first with your horse consultant and then with a solicitor. Unfortunately, just because a seller refuses to warrant a horse does not always mean it has a fault and may simply indicate a super-cautious seller, in view of the current legal position of sellers and the expense of court cases. You may simply be offered a Bill of Sale, which states the horse is sold as seen, tried and approved, putting the onus firmly on the buyer.

A half-hour consultation with a solicitor should give you clear advice as to what to aim for and what recourse you may have if things later go wrong, so it is well worth having, I feel. Any legal practice should be able to help.

Insurance broker

Many people don't bother to insure their horses, and with good reason. As one who has in the past been on the sticky end of dealings with insurance companies who refused to pay out on a perfectly legitimate claim, I can well understand people's reluctance to insure their horses. It should be remembered that insurers are gamblers and they only bet on near dead certs. If they think you and/or your horse are very risky they shorten the odds and say they may be able to insure you, but that it will cost you (in other words, it will cost you a lot and possibly more than you feel worthwhile). They may even refuse to insure the horse at all for some reason, or put so many exclusions into your policy that, again, it is not worth taking out. All insurance premiums seem expensive and the more activities you intend to pursue with your horse, and the more active and 'dangerous' – in their eyes – is the pursuit the more they will charge. You can cover the horse for death by accident, illness or humane destruction, or for loss of use due to disability when you should get a stated percentage of its former value.

Unfortunately, although this sounds cut and dried, it may not be so in practice. I know several people apart from myself who have claimed loss of use, only to find that the insurance company's vet disagreed with their own as to the horse's capabilities, and they were asked to make the horse do, or try to do, what it was no longer capable of, thereby causing it considerable distress amounting even to cruelty. Right-thinking owners who refused to treat their horses so abominably were refused payment on loss of use, and two I know who did push the horse to prove his disability were still not paid due to the veterinary surgeon's disagreement. One owner was subsequently brought to court by an onlooker for causing her horse unnecessary suffering but was acquitted

– apparently the court decided that claiming money was more important than safeguarding the horse's wellbeing!

If you need to have your horse put down suddenly because say, it breaks a leg in the field, you may feel that getting your vet to contact the insurance company stating the reasons for the destruction would be sufficient to ensure your payout. Not always so! It is certainly not unknown for a horse to have been destroyed for humane reasons (to prevent further suffering) and for the company then to refuse to pay out because its own vet was not first consulted. In one case, the company really did expect an owner to keep her horse alive with a broken leg over the weekend till it could be contacted and get its own vet over to see it on Monday morning. Because she didn't, she didn't get her money. Not all companies operate such appalling policies but some certainly do, so it is essential to have a detailed discussion with a broker and written confirmation of what the small print really means.

There are two types of insurance which are needed: third party to cover any damage your horse may do to other people or their property; and cover for vet's fees in the case of major illness or accident as some kind of guarantee (as far as you can secure one) that in such an event you *will* have the money to cope.

Third party insurance is free for members of the British Horse Society and Pony Club as part of normal membership; and as the subscription is cheaper than the premium for separate insurance you would be well advised to join for this reason alone, in addition to the many other benefits.

Cover for vet's fees is normally quite reasonable compared with a premium for the value of the horse himself and is worth looking into, but do ask about future exclusions and possible premium rises *after* a claim. For instance, you could claim for treatment for a sprained tendon, but when you come to renew, find that either the premium has risen alarmingly or that all future leg problems are excluded. Get anything you are uncertain about clearly explained in writing, including an explanation of ambiguous or unintelligible small print, and ask a broker or solicitor for advice if you are still unclear.

Some companies give discounts for horses kept on clean air régimes (dust-free bedding and fodder) because their susceptibility to disease, and especially respiratory disorders, is lower; and some give discounts to BHS members, so ask about this, too. Ask about all discounts in a blanket way such as: 'What premium discounts do you give?' You may find you get a discount for some quite obscure reason such as membership of Bodgeville Morris Dancers' Club.

Crib-biting (above), box-walking and weaving (below) are all stable vices which result from horses being over-confined in too-small stables and not given enough exercise, particularly at liberty, or from some other intolerable stress in the horse's life. The solution to the problem lies not in physically preventing the horse from performing his vices, which at least give him some release from the stress and tension, but, by careful analysis and observation of the horse and his lifestyle, to find out why the horse is distressed and to do all you can to remove the cause. However, in confirmed cases of any vice it may only be possible to lessen it and not to prevent it altogether. Adequate liberty lessens most vices and can stop them developing.

(aggressive!) at feed times. Little things like this may not bother you, but something might surface which does, and it's as well to know about it before you go to any expense, for example of insuring him for a week's trial, if one is permitted.

If by this time you really feel you want him and your adviser approves him, by all means now say you'll buy him subject to a satisfactory veterinary report (and a weeks' trial, if applicable). It may be better not to put down a deposit at this stage unless dealing with a really reputable professional vendor, as there might be a difficulty in getting it back, should the horse not pass the vet – although in theory there shouldn't be.

Contact the vet you've chosen and arrange for him to make his (or her) own appointment to examine the horse. You need not attend, unless you particularly want to. The vet will send you a report on the results and advise whether or not the horse is suitable for your purposes. If there's anything to discuss in addition he or she may well ring you.

Veterinary certificates of soundness have already been mentioned, but it is as well to note that for the standard veterinary certificate things like the height of the horse, temperament, freedom from vices, or whether drugs have been administered prior to examination (such as pain-masking drugs to hide lameness) are not the responsibility of the veterinary surgeon; these should be included in a separate written warranty, which is the vendor's responsibility. If you want things like blood tests to detect drugs, leg x-rays to detect early bone disease (not always reliable) or other medical tests you will have to tell the vet and vendor beforehand. It's best to be guided by the vet and possibly your adviser on this if there is anything worrying you.

If you want a horse of a specific breed, one registered with a specific organisation or something similar, you will also have to state that when the horse is handed over to you you will also want his pedigree, registration papers (all of them), passport if applicable, and vaccination certificates. If such things are really important to you, I would advise you not to hand over your money till you have them, as such things have an unpleasant habit of never materialising after change of ownership – a cynical comment but one borne of experience.

If things somehow get delayed and the vendor starts putting pressure on you to buy with remarks like: 'I have someone else interested in him and they want him quickly. Do you want him or not?' ask your adviser to intervene. It may be true, but it could be a ploy to get you moving when perhaps you yourself are not certain and are viewing other horses. You'll have to decide whether or not you really do want the horse – but if you *are* prevaricating, perhaps, subconsciously, you don't. Only you can decide! If you are certain and are simply waiting for insurance and transport arrangements to gel, put a deposit on the horse (10 per cent is normal) and get a written receipt stating which horse it reserves. If he's that nice he may well go elsewhere if you don't act.

On the subject of money, it is not the vet's job to give an opinion on price, or as to whether the horse is fairly valued; this is your adviser's responsibility, so take her advice. You don't have to follow it – and if the horse really *is* your dream horse you may well feel justified in paying over the odds for him. Dreams don't often come true in life.

On the other hand, perhaps it's prudent not to wear your heart on your sleeve and let the vendor know that you have actually fallen in love at first sight. He or she could easily come back to you before the sale has taken place and say someone else has offered £X more for the horse, justifiably believing that you'll stump up the extra. If you play it cool you stand more chance of getting a better deal; though most vendors are perfectly honest people, professionals have to make a profit to live. I personally dislike too much unreasonable haggling. It's insulting to a fair vendor with a good horse fairly priced, and horses in that category are easy to sell – so again, if you antagonise the vendor or play hard to get you may lose the horse.

T he horse family is unique in the animal world, with very specific specialised physical and mental functions which mesh together to provide the ideal survival system for the horse in its natural condition – a grazing and browsing lifestyle in an open environment where predators abound but with plenty of escape routes plus safety among the numbers of the herd. These natural conditions produced a animal which needs particular management techniques to remain healthy and content.

DIGESTION

It is amazing that horses actually manage to stay alive on what they eat when you look at the way in which their bowel is arranged. Equine animals, unlike cows and sheep, do not have a rumen in which fibrous food is processed, nor do they 'chew the cud'. Their food is digested partly by the action of chemical enzymes in the stomach and small intestine, and partly by bacterial action in the caecum. Their gastric secretions are more or less continuous and are hardly influenced by the presence of food, not like humans or carnivores whose gastric juices are stimulated by the sight or smell of food.

Horses out at grass need to eat for most of the time, and when being fed artificially they require small meals at fairly frequent intervals. The bacteria responsible for digestion in the caecum are sensitive to changes in the diet, and as a general rule it is important to feed a varied diet so that the bacteria do not lose the ability to digest a particular component. It is not a good idea to give oats one day and bran the next, for example – all components of the mix you use should be given at each feed. Compared with the rumen of a cow the horse's caecum is relatively small, and the process of bacterial digestion has to be correspondingly very efficient; even small disturbances affecting the efficiency of the process can have major consequences. The products of bacterial digestion are produced in the caecum, which lies at the lower end of the gut; this is after the food has passed

by Simon Wolfensohn, MA, Vet MB, MRCVS

2 IN HEALTH AND DISEASE

through the small intestine (which is where absorbtion of nutrients mainly takes place in most animals) so there is not much of the bowel's length left from which these products can be absorbed. The horse is therefore very easily affected by periods of even relative starvation, and any horse expected to work hard requires a good supply of high quality feed to compensate for the lack of time spent grazing. Finally, every horse or pony of course needs a constant supply of good, clean water and must have regular opportunity to drink.

THE HEART AND LUNGS

Horses are athletic animals, built for running, and their cardiovascular system has to be efficient. The heart and lungs are essentially the same as those of most mammals, but it is important to have a basic understanding of the way they work in order to understand the veterinary problems that may arise.

ANATOMY

An illustration showing the points of the horse appears on pages 8–9 and diagrams of the foot and the digestive system appear on pages 76 and 96 respectively. The horse is adapted for running and grazing and the main features of the anatomy which differentiate it from other mammals are principally to do with these adaptations.

Working our way from head to tail, the head is equipped with jaws designed to grind food as a preparation for digestion in the stomach, with the front teeth (incisors) arranged to nibble and pluck grass. The eyes are set high up on the bone of the head, which needs to be relatively massive in order to accommodate the muscles needed to move the heavy jaw bones. The ears are also set high to optimise their position for sound gathering.

The neck is long, as you would expect in a grazing animal, but its length is also an asset when the animal is watchful and the head is raised. To support the muscles needed to carry the head the bones of the neck are quite large. The bones of the spine (the vertebrae) are massive, to accommodate the muscles of the trunk and to support the muscles of the back itself, which forms the girder from which the limbs are slung. The spinal column also houses and protects the spinal cord. The back is a massive structure which makes it ideal for our human purposes as a load carrier or for riding, and in natural conditions it is well protected against injury. It is only when we are asking an unfit or untrained horse to expose its back to an unnatural loading that we might expect injuries to occur.

The limbs are long and the skeleton is adapted so that the horse walks on the last phalanx of his single toe in order to increase the length of the leg. The feet themselves have become very specialised, to bear the weight needed and to cushion themselves from injury. Two structures in particular are important in taking the load imposed on the foot: the sensitive laminae which help to distribute weight, and the digital cushion. To understand how the laminae work, imagine trying to push a cone of wood into a bucket of sand: the sand will support the pressure over the whole surface of the cone. But if you were to push the same cone into an empty bucket, all the pressure would be concentrated on the point of the cone. The sensitive laminae work in the same way as the sand to spread the load.

The digital cushion lies under the frog and is made of fibro-elastic tissue. It acts as a shock absorber by virtue of its elasticity, and it is thought that it also acts as a pump, helping the blood to return up the leg. The frog itself is made up of intertubular horn (ie horn which does not contain the horn tubules which are the strengthening structures of the hoof horn) and is constantly growing. It helps to prevent the foot slipping and distributes pressure; a healthy frog is essential for normal foot function. The frog, sole and hoof wall all grow downward at the same rate.

Another interesting feature of equine anatomy is the 'stay apparatus', an arrangement of tendons and muscles giving automatic support for the hind limbs, which allows the horse to remain standing for very long periods without significant muscular effort. The most significant feature of this apparatus (also sometimes called the reciprocal apparatus of the hind limb) is that the hock cannot flex or extend without the stifle. The tendons of the front limb also carry out this function to some extent.

IN HEALTH AND DISEASE

The heart and lungs
Anatomy
What makes a good horse?

WHAT MAKES A GOOD HORSE?

When you come to buy your first horse you need to have some idea of what constitutes a sound conformation. What features of the body and limbs are particularly important in making a horse whose athletic abilities will be up to scratch?

The back should be fairly short and strong, especially in draught horses, although Thoroughbreds need a slightly longer back; however, all types should have a relatively short lumbar area of the spine. The middle of the back should curve very slightly towards the ground – a completely straight back is a fault. This curvature will increase naturally as the horse ages, but an excessive curve is a fault in a young horse, and is referred to as a dipped or hollow back. Equally the back should not curve upwards, a fault known as roach back. Although a roach back does not appear to affect a horse's performance it looks bad and is uncomfortable to ride. Draught horses and jumping horses have a more sloping rump, often called a jumping rump, but this should not be confused with a goose rump which is a sloping but poorly muscled type of conformation.

Turning to the legs, the horse should be generally well proportioned and balanced. Starting at the front, the shoulder should be long with a reasonable slope to it; if too upright the gait may be shortened and the horse may stand over in front ie tend to lean over its own forefeet. The elbows should not turn in or out. The bone of the lower leg should be sufficiently developed to carry the weight which might be expected of the horse, but in good proportion to the rest of the leg. The characteristics of the parts of the lower leg will vary depending on the type of work expected of the horse.

The knee should be large and straight to avoid any tendency to over-extension of the joint. Similarly any forward bowing of the knee (over at the knee) should be avoided, and lateral deviation of the knee joint is a serious fault. The fetlock and cannon should be quite straight when seen from the front. Opinions vary about the length and slope of pasterns but in general you should look for a moderate length and slope. Each foot should match its opposite number. The more even the surface of the hoof, the better, and the frog should be well developed. The sole should be slightly concave. The heels should be well shaped and the axis of the front foot (the angle between the slope of the foot and the ground surface) should be approximately 45–50 degrees. The hind foot axis should be 50–55 degrees. For most purposes it should be possible to continue this foot axis in a straight line

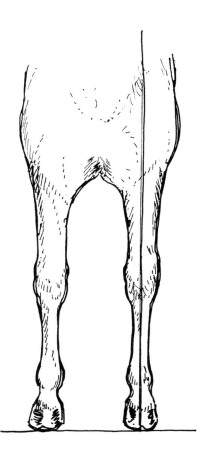

When viewed from the front, the horse's hoof should be directly under his point of shoulder and the hoof facing straight to the front to ensure natural and direct forces up and down his leg, a conformation which enables the horse to withstand the impact of movement with minimum chance of injury.

The conformation of the feet and legs is of particular importance, especially the hoof/pastern axis. The hoof wall and coronet should be the same height at both sides when viewed from the front, as here (far left) and the angle the hoof wall at the toe, and the long pastern bone, form with the ground should be about 45° for the front feet although a slightly higher angle (about 50°) is often acceptable for the hind feet. A point often overlooked is that the angle of the heel wall should be the same, and also that of the shoulder, when considering the forehand.

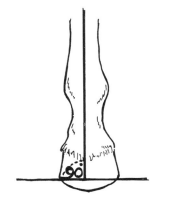

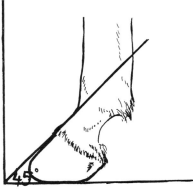

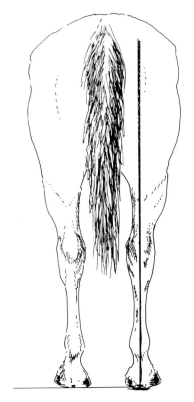

From the back, you should be able to 'see' a straight line dropped from the point of the buttock, through the hock, down through the centre of the cannon and fetlock and directly through the heels to the ground.

Check legs for lumps, wounds and heat every day. Early treatment can prevent deterioration to a more serious condition later.

through the pastern. The feet should be straight, and there should be no sign of either pigeon-toes or the opposite, splayed feet.

As far as the hind leg is concerned, it is important that the horse is well muscled as the hind quarters are his propulsion unit. There is a wonderful expression quoted by Horace Hayes in his *Veterinary Notes for Horse Owners* that a horse should have 'a head like a duchess and a bottom like a cook'.

Hocks should be straight when you look at the horse from behind. Cow-hocks, where the hocks turn inwards, or bowed hocks – turning outwards – are both faults. Nor should there be excessive angulation of the hock, 'sickle hocks', when seen from the side; at the other extreme the hock should not be too straight. The horse should stand with the feet well placed vertically below the hip, not backward in a rocking horse type of stance or too far forward under the body.

All these conformational faults have the effect of putting extra strain on joints, tendons and ligaments, especially when subjected to hard work. They unbalance the horse's action and produce abnormalities of gait which in many cases may be more of a clue to any problem than a mild conformational weakness itself. However, problems of this sort are not easy to spot without practice and you should take an experienced horseman or horsewoman with you when you first go to look at a prospective purchase. Finally, a veterinary examination for soundness, which is a pretty exhaustive procedure, is the only real guarantee that you are buying a horse which, as far as anyone can discover, is genuinely sound.

THE HORSE'S HEALTH

In order to know when a horse is unwell you first need to know what he should look like when everything is normal. Begin by looking at him over the stable door and note his breathing, his shape and the condition of his coat. The normal respiratory rate of a resting horse is around fifteen breaths per minute, and breathing should be almost silent. As the horse exhales you should see very little movement of the abdomen. There should be no nasal discharge other than a little clear mucus and there should be no coughing, although of course like any other animal a horse may cough occasionally if there is dust in the air. Some horses habitually cough and snort as soon as you put a saddle on them.

Look at your horse often at rest so that you know his or her normal shape. Is there any sign of him being more tucked up than usual? Is there any distension of the abdomen? Don't forget that in the case of a mare the most likely reason for enlargement of the abdomen is pregnancy. Horses with broken wind often show a distinct 'heave line', along the abdomen about a fifth of the way up from the lowest point. This is where the abdominal muscles which lie along the length of the belly, and which have become more developed because of the effort of breathing, join the oblique muscles of the trunk.

Your horse's coat should be smooth and glossy with no bald or thin patches. The ribs should be faintly visible below the muscles and when the horse is hot you expect to see the blood vessels of the skin quite prominently. A resting horse should not sweat and there should be no swellings in the skin.

Take note, too, of your horse's behaviour. Out in the field a sick horse may detach himself from a group and stand alone. He may spend more time lying down than usual. He may adopt an abnormal position, pointing one forefoot forwards, for example, because of pain in the foot, or standing in an exaggerated rocking horse type of position with an acute laminitis. In the stable you may notice him being restless or pawing at the ground. He may turn himself round more than usual, and he may turn round to look at his flank if in pain. In the stable or outdoors he may try to roll. Look out for any unusual behaviour in feeding or drinking, and watch out for salivation or drooling, which should not normally occur.

The eyes should be clear, bright and open, without any redness of the lids or sclera (the white of the eye). If the horse has been out at grass there may be a slight yellowish tinge to the sclera. There may be a little clear fluid or dried brownish discharge at the corners of the eyes. However, any swelling of the lids is abnormal. The two eyes should match each other (except for the colour of the iris in some cases). In a dark stable if you shine a small pen torch into the eye you should be able to see the pupil contract. Any other tests of vision are very difficult to carry out and are probably best left to your vet.

The ears should be clean, responsive to sounds, and mobile. When you handle them there should be no sign of pain.

The lips of a normal horse are clean and dry and there is usually only a slight sweet smell on the breath, although you may recognise a distinctive smell if the horse has been eating molasses for example. Any foul or offensive smell is abnormal. Normally where there is no dark pigmentation the gums are pink, although you may notice paler areas, areas with a faint blue tinge or with a yellow tint (in horses at grass). If you press the gum firmly with your finger it should whiten as the tiny capillary blood vessels in the surface empty and the colour should return in a couple of seconds (this is referred to as the capillary refill time). Next look at the teeth, and look out for food particles trapped around them, and for sharp edges or hooks – however, don't forget that hooks on the first molar teeth are quite normal at seven or eleven years old. Take great care when examining the mouth of a horse, and never try to put your hand inside or you may lose your fingers! Watch the horse as he eats, and look out for any difficulty in picking up the food, in chewing ('quidding') or in swallowing.

Look for any swelling under the angle of the jaw at the top of the throat, which is where the signs of strangles first appear. The veins of the neck are usually barely noticeable so any apparent distension is significant, although in thin skinned horses you may be able to see a 'jugular pulse' when they are standing quietly at rest.

As you look along the back the two sides should be evenly matched, there should be no swellings, and if you run your hand firmly along the back there should be no signs of pain. When you lift the tail the anus should be clean, without any discharge and the vulva of a mare should also be clean, although there might be a very little clear mucus or dried brownish discharge. In season there will be a fairly profuse, clear mucus and the lips of the vulva may be mobile. Look under the tail and check for any swellings, especially in grey horses where this is a common site for a type of tumour referred to as a melanoma.

A normal horse should have clean limbs free of blemishes or swellings, and you should be able to flex and extend each joint through a full range of movement. (Testing the joints is something you need to be shown how to do by your vet. It is not safe for you or the horse to do this without instruction.)

The hoof is normally smooth and should be kept well trimmed or shod, and be dressed regularly with hoof oil. Look out for any splits or cracks which may need attention, and check for loose nails. Look carefully at the coronary band at the top of the hoof for any injuries, as any damage here is potentially serious. The frog should be supple and well shaped, and the angle where the hoof wall joins the frog should be clean, dry and firm. The sole should be firm with no discoloured areas, no crumbling areas and no painful spots.

Before riding you should walk your horse enough to check for any lameness or stiffness; and before you set off you should look at the stable to make sure that any dung is normal, that the food has been eaten, and that there is no sign of any abnormal discharge on the floor or walls, nor signs of disturbance to the bedding due to pawing or rolling.

When riding you should watch out constantly for any sign of lameness, any increase in respiratory effort, abnormal tiredness or signs of discomfort.

Get to know your horse really well so that you can spot abnormalities quickly.

IN HEALTH AND DISEASE
The horses's health
Abnormalities of gait

ABNORMALITIES OF GAIT: THE HORSE'S ACTION

A horse should have a level action when seen moving on good ground. The head should be carried straight and level, and the back should remain level as he moves; there should be no uneven movement or rolling motion in the gait. Each foot should move cleanly without interfering with the movement of the other feet, and there should be an impression of controlled, positive movement. However, it is important to distinguish abnormalities of gait from the clumsiness or inattention which you might see in a young untrained horse or an unfit horse. Abnormal gaits include:

Brushing — the fetlock is struck by the shoe of the other foot as it comes through. In most cases brushing is more likely to be due to poor condition or lack of training than conformation.

Dishing or *Paddling* — outward deviation during the flight of the foot, landing on the outer edge. Often due to pigeon toes.

Winging — the foot moves too far towards the centre as it travels, landing on the inside of the foot. Most common in horses with a wide-based stance and toes turned outwards.

Speedy cutting — happens at faster paces, especially when the horse is tired. The inside of the lower part of the knee or hock is struck by the shoe of the opposite foot.

Over-reaching — the shoe of the hind foot catches the heel of the fore foot, generally at a gallop or in jumping. It is more often due to over-extending the horse than to conformation.

Plaiting — the foot flight arcs inward and lands in front of the opposite foot. Seen in horses with a narrow stance and turned out toes.

Just a word of warning on the diagnosis of abnormalities of gait or lameness: don't draw conclusions from examining a horse on an uneven or slippery surface; and make sure that whoever is leading the horse walks it calmly and trots it so that the head has freedom of movement. It is very easy to produce an abnormal gait by bad handling.

HEALTH MAINTENANCE

Preventive medicine and routine veterinary care are vital if you are to ensure the good health of your horse and avoid unnecessary worry and unnecessary problems. Most preventive medicine really comes under the heading of good management – making sure that stables are well ventilated, that feet are cared for regularly, that horses are fed and watered correctly, and kept in suitable fields with proper fences (*not* barbed wire). If more horses were managed and looked after properly the veterinary profession would have a lot less to do!

There are a few specifically veterinary matters that need attention in the routine health care of your horse: worming, vaccinations and dentistry.

Worming

Good pasture land is often in short supply, and as a result quite large groups of horses are often out at grass together. Contamination of pasture with worms often becomes heavy, and so regular worming is very important both to avoid damage to the individual horse and to reduce the contamination of the pasture.

A worming programme has to deal with round-worms, tapeworms and preferably also the larvae of the bot-fly (Gastrophilus) which are found in the stomach of horses. It is important to take the advice of your vet on which wormers to use; worms may develop resistance to one particular drug (especially thiabendazole and some related compounds) and it may be advisable to vary the wormers you use to avoid this, although some of the more recent (and more expensive) ones do not suffer from this deficiency. Some such wormers, however, are not effective against tapeworm eg ivermectin and others only have a limited effect against them – fenbendazole is one.

Bot-flies lay their eggs on the horse's coat, and these are then licked off and swallowed. In the stomach they develop into larvae which attach to the stomach wall and are passed in the faeces the following spring. The larvae are thought to cause little damage, although large numbers may interfere with digestion, but the adult flies can be a major nuisance and dosing for bots will help to break the life cycle and reduce the numbers of flies.

Tapeworm (principally Anoplocephala) can affect the horses by interfering with digestion, and in large numbers they can obstruct the passage of food through the bowel. Quite recently it has also been recognised that they can be a cause of colic, and they have therefore taken on rather more significance than in the past. Any effective worming scheme should certainly include tapeworm.

Roundworms of various types are the most important species which we have to control. Foals are affected by two main species: ascarid worms (Parascaris equorum) and threadworms (Strongyloides westeri). The larvae of Parascaris worms enter the body through the gut and cause damage to the liver and lungs. These worms can cause severe illness by blocking the gut or by causing peritonitis, or a severe cough with secondary complications. A foal infected with these worms looks pot-bellied, undersized and in poor general condition. To prevent infestation the mare should be dosed before foaling, dung should be picked up daily, and stables and paddocks should be kept as clean as possible.

Strongyloides is a much smaller worm which only causes serious disease in the very young foal. The infective larvae of the worm reach the foal via the mother's milk, and can cause severe diarrhoea and sometimes death.

Older horses may be infected with large red worms (Strongylus or Triodontophorus), small red worms (especially Trichonema and other small strongyle-type worms), pin-worms – also known as seatworms – (Oxyuris equi) and lungworm (Dictyocaulus arnfieldi).

Of these, Strongylus vulgaris infection is the most serious, and it is thought to be the most common cause of bouts of spasmodic colic. One horse can contaminate pasture with this worm at the rate of thirty million eggs per day. The horse becomes infected by eating infective larvae on the grass, and these then pass into the gut and enter the bloodstream. Damage to the blood vessel walls causes blood clotting and blocks the circulation to the bowel; this may cause part of the bowel to be destroyed, which kills the horse. In other cases the larvae pass back into the gut, mature and damage the lining of the bowel. Some horses with this type of worm may show no symptoms at all, others may be in poor condition, and others may suffer colic or sudden death.

Small red worms enter the bowel with the food and then grow to maturity, attaching themselves to the wall of the bowel. Symptoms which result are poor condition and anaemia; some horses may develop diarrhoea, and others constipation, depending on the level of infection. Routine dosing with wormers may not get rid of this worm because it buries itself in nodules in the bowel wall; when infection is suspected or proven it may be necessary to use higher dose rates of wormers (but only under veterinary supervision).

Oxyuris is not a very significant infection. The worms cause irritation around the anus and may cause intense rubbing and hair loss, but routine doses of most wormers will get rid of the infection. The eggs of the worm can often be seen around the anus when infection is present, so diagnosis is usually easy.

Lungworm (Dictyocaulus) is mainly a parasite of donkeys and many donkeys can be heavily infected with them and show no symptoms. In the horse the infection causes coughing. Diagnosis can be difficult because only a few horses will pass the larvae in their droppings (less than one in ten). Several modern worm drugs are quite effective against lungworm and it is not as much of a problem as it used to be.

Before turning any new horse out to grass it is advisable to give an effective worm dose and to keep it stabled for twenty-four hours in order to avoid potential pasture contamination.

Foals should normally be wormed from 6 weeks of age, unless diarrhoea due to Strongyloides infection requires treatment sooner. Dosing should then continue every 4–6 weeks. Adult horses should be dosed at the beginning of March and if they have been grazing over the winter they should be dosed at a level adequate to get rid of larval worms, or you should use a wormer which kills larvae at normal dose rates (eg ivermectin). Dosing should continue every 4–6 weeks until early October when another larvicidal dose should be given. A wormer effective against bots should be given in November or December, and routine treatment should occasionally include a wormer active against tapeworm eg pyrantel (Strongid).

Vaccinations

All horses should be vaccinated against tetanus and equine flu. In special circumstances there are also other diseases for which vaccinations may be given, for example equine rhinopneumonitis.

Tetanus: Horses are very susceptible to tetanus, and infected wounds such as nail punctures are particularly dangerous. Infection can follow even superficial skin wounds or surgical operations, and unvaccinated horses should then be given tetanus antitoxin to prevent infection. Many vets prefer to give antitoxin even to vaccinated horses.

Most vaccine manufacturers recommend revaccination every eighteen months or two years after an initial course of two injections; this could start at three to five months of age, and a first booster dose given after one year, but in many cases vets will give an annual booster for tetanus. Pregnant mares should have a booster vaccination one month before foaling. The bacteria which cause tetanus are found in soil and the risk of tetanus varies from one area of the country to another depending on the type of soil. The bacteria responsible, Clostridium tetani, can also be found in horse droppings. The signs of tetanus usually develop weeks or even months after the initial infection; they begin with increasing stiffness and difficulty in walking, progressing to convulsions and death. There is no effective treatment in most cases, and the majority of horses which contract tetanus will die from it.

Equine flu: There are two main strains of the flu virus: the Prague strain and the Miami strain. It is a very contagious disease and epidemics are common; horses remain infective for about a week, and the disease has an incubation period of two to four days. It is most common in groups of young horses, although any susceptible horse may contract the disease from an infected horse.

Vaccinations usually start at three months, with a second injection after four to six weeks. The first two booster injections are then normally given at six month intervals and revaccination after that is usually carried out every nine months to a year. Unfortunately, flu vaccination has gone through periods of having a bad reputation because horse owners have found that their horses sometimes still suffer from flu-like symptoms, or have developed an adverse reaction to the vaccination. It must be remembered that there are many other possible causes of respiratory infection in horses – also that a sick horse may be suffering from the effects of more than one respiratory infection at once, so flu vaccination may sometimes appear to have been ineffective, although this is not really the case. Because flu is such a contagious disease and is largely preventable it is still very important that as many horses as possible are vaccinated against it. Reactions to vaccination often consist of swellings around the area where the injection was given, but they are not usually serious. A few horses may develop a mild cough and be rather off colour for a few days after vaccination; a very few may develop more serious reactions. It is important to remember that the number of horses which show reactions to vaccination is comparatively very small compared with the number being vaccinated. Horses should rest for about seven days following vaccinations.

Flu usually begins quite suddenly with a fever, a deep dry cough, and a nasal discharge which may become very mucky if secondary infection occurs. An affected horse should be rested and isolated if possible. Treatment includes antibiotics to control any secondary infection, and good nursing care which should protect against dusty atmospheres and draughts, provide palatable food and drink, and keep the stable thoroughly clean, warm and dry. To keep the horse comfortable it is also necessary to bathe away any discharges and groom the horse thoroughly but gently. In some cases inhalants might be used to relieve the respiratory congestion, or your vet might prescribe a drug to help keep the discharges as fluid as possible. After recovery your horse should continue to rest for around ten days before getting back into work. Flu can have lasting effects on a horse's wind so should never be neglected.

Dental care

Regular dental attention is essential to maintain good health. Many horses, particularly older animals, will develop a slight misalignment of the large molar teeth (the cheek teeth) so that a sharp line of tooth enamel is left rubbing against the inside of the cheek; a corresponding overlap on the inside of the molar teeth may injure the tongue. Horses affected in this way may have difficulty chewing, which is often spoken of as 'quidding'. This condition is easily dealt with by filing the sharp edges off with a rasp, also referred to as a 'float' by some people. Some vets prefer to use a mouth gag of one kind or another to keep the jaws open, others prefer to work without one.

Not much fun, but rasping (smoothing) teeth is sometimes necessary.

Although there are many lay people who are quite competent at rasping teeth, the proper person to carry out this job is your veterinary surgeon. For one thing it might be necessary to give a sedative in order to rasp the teeth properly and avoid stress to the horse, but also it is a good opportunity to examine the mouth thoroughly for other signs of disease – often in the course of routine rasping a vet will find other problems which need attention. These might include dental decay, tooth root abscesses or problems because of wolf teeth (the supernumerary and vestigial first molars). By seeing early signs of trouble it is often possible to avoid anything more serious in the future.

It may be convenient for the vet to rasp your horse's teeth for you at the same time as giving booster vaccinations, so it is much better to arrange to have the horse in the yard where clean water is available and where suitable restraint is possible, rather than out in the field.

HEALTH PROBLEMS
Some of the more common diseases

Abortion: is the loss of a foal before the three hundredth day of pregnancy, and can be caused by bacterial infection, fungal infection or viral infection, in particular with equine herpes virus. There are also non-infectious causes including the presence of twins, abnormal foetal development or poor nutrition. Serious injuries can also lead to the loss of a foal. Equine herpes virus is infectious and is the agent responsible for 'abortion storms' or epidemics, so if a mare aborts it is important for your vet to have the opportunity to make a precise diagnosis – she should be isolated until the cause is known.

Bad back: There are several conditions which may affect the muscles, ligaments and bones of the back. Specific conditions include damage to the dorsal spines of the vertebrae, which may touch each other and cause inflammation when the back is flexed, and arthritis of the lumbar spine, seen in older horses. Disc injuries and fractures are rare. Muscle injuries and bruising may occur due to bad riding or over-exertion.

Bronchopneumonia: Not a specific disease in itself as it may be caused by a variety of bacteria, and is often secondary to an initial virus infection such as flu. It is more common in foals especially if kept in poor conditions or deprived of colostrum from the mother after foaling. Signs of bronchopneumonia include rapid breathing, progessive difficulty in breathing, lack of appetite and dullness. Treatment consists principally of antibiotics, expectorants and drugs to liquefy secretions, together with good nursing care as described for flu (in the section on vaccinations).

Choke: Choke means an obstruction of the oesophagus between the mouth and the stomach. It may be caused, for example, if a racehorse is given dry hay too soon after racing, while still rather dehydrated – the horse is unable to swallow the food properly and it becomes lodged in the oesophagus. Sugar beet pulp which swells when mixed with saliva, or a specific object such as a small ball, are also potential causes of choke.

Signs of the condition are restlessness, refusal to eat or drink, or possibly attempts to do so but with the food or water being regurgitated at the nostrils. It may be possible to feel the obstruction.

Many cases of choke will improve on their own as the food softens and digestion begins, but other cases may require veterinary treatment. Often this involves sedation and the passage of a stomach tube to allow the material to be flushed away; some cases may even require surgery. It is always best to call your vet to assess the situation rather than waiting to see what happens, as delays may result in dehydration or serious metabolic disturbances.

COPD – (Chronic obstructive pulmonary disease) is a common cause of coughing in stabled horses. It is an allergic response, usually to the fungal spores from mouldy hay or straw, and if left untreated or allowed to become severe this condition leads to 'broken wind', which is also called 'heaves'. The allergic response produces mucus and contraction of the small airways of the lung, and this obstruction can then result in distension and rupture of the small air sacs, which are the termination of the airways (this is called emphysema). The signs of the condition are very variable depending on the degree of obstruction, and vary from a mild cough to severe difficulty in breathing. Diagnosis is based on response to removal of potential causes of the allergy from the stable and surrounding area; response to treatment; or endoscopy, which involves passing a flexible tube with a lens and light source into the trachea where excess mucus can then be seen.

Other conditions with which COPD can be confused include lungworm, and bronchitis due to infection. If diagnosed early, before the lung is damaged, the condition can be well controlled using antihistamines and drugs (bronchodilators) to dilate the small air passages. It is also necessary to avoid hay and straw as far as possible, give well-dampened hay or hayage, and coarse mix, cubes or thoroughly damped feeds, and ensure good ventilation of the stable. When turned out to grass mildly affected horses will improve almost immediately. Once emphysema has developed the changes are permanent and the horse will remain broken-winded and incapable of hard work.

Colic: The word 'colic' is applied to any pain in the abdomen, but in common usage it generally refers to pain originating in the gut. Horses seem to have a very sensitive bowel, especially Arab horses, and the pain of colic is often intense. Signs begin with restlessness, lack of appetite and sweating, and as the pain gets worse you may see the horse looking round at his flank, pawing the ground or stamping his feet, rolling, kicking at his abdomen, groaning or even bellowing.

Colic is quite common and is due to a wide variety of causes. Many minor cases improve spontaneously or with

When your horse gets up from rolling check that he has a good shake, like this. Rolling without shaking off the dust afterwards can be a sign of internal discomfort or pain, such as in colic.

minor veterinary treatment; other more serious cases may require surgery, and it is essential that the decision to operate is taken quite early in order to have a reasonable hope of success. Therefore any case of colic should be treated as an emergency. However, it does unfortunately remain true that some cases of colic will end in death despite surgery, and that sometimes humane destruction of the horse is the only possible course of action. Types of colic include:

Spasmodic: overactivity or spasm of the bowel wall causes pain which usually responds quite quickly to painkillers. This is probably the commonest type of colic and may be caused by bad feeding or watering in many cases. Failing to take the chill off the horse's water is a common cause.

Flatulent: due to distension of the bowel with gas; this usually follows feeding with fermentable foodstuffs.

Impaction: usually affects the colon or caecum and is due to poor gut motility, dehydration or eating food contaminated with sand or grit. Most impactions respond to treatment with lubricants such as liquid paraffin.

Obstruction: this is due to the bowel twisting on itself in various ways and causes very severe pain, shock and ultimately death. Some cases may be suitable for surgery, but others may be hopeless.

Thromboembolism: caused by worm damage affecting the blood supply to the bowel. In severe cases a large section of bowel may be damaged and the outlook for the horse is hopeless; in others only a small part of the circulation may be involved and other blood vessels compensate for the damage.

Idiopathic: some cases of pain have no discoverable cause and these are called idiopathic.

Your vet will need to take a large number of factors into account in assessing whether a case of colic is going to need surgery, and any information you can supply through careful observation of the course of the condition, of the feeding history and watering will all be useful. Your vet will look at the pulse rate, respiration rate, and the mucous membranes, he will listen to the sounds from the gut, possibly take blood samples, or carry out an examination by feeling the bowel through the rectum via the anus, and he may take a fluid sample from the abdomen. He will also take into account the response to initial treatment and other factors before making an assessment.

It is now generally agreed that walking a horse with colic is of no particular value, but it is probably wise to try and prevent him rolling whether outside or in the stable. Horses with colic can be very dangerous to the handlers, however, and experienced help should be obtained as soon as possible as well as calling the vet.

Conjunctivitis: is the inflammation of the tissue lining the eyelids. It may be due to infection or the presence of foreign material such as a piece of straw. The eye will look red and sore and there may be a mucous or purulent discharge. First-aid treatment consists of bathing the eye with saline solution or a proprietary human eye wash, but because of the possibility of a foreign body it is advisable to get prompt veterinary attention. Conjunctivitis may also be seen in the early stages of flu infection or other infections such as strangles.

Gently open the eyelids with thumb and forefinger as shown, carefully squeezing in ointment without actually touching the eye.

Contagious equine metritis: This disease is a bacterial infection which first occurred in Britain as an epidemic in 1977 and has been found in many other countries as well. Infection is introduced by an infected stallion covering a mare which will then develop infection of the uterus (metritis); a purulent discharge from the vulva is usually the visible sign. After recovery the mare may remain a carrier of the bacteria which cause the disease, and she may transmit the infection to another stallion.

The disease can be prevented by routine bacteriological swabs taken from the mare and stallion before mating.

Cough: The most important causes of coughing of an infectious nature are flu and equine rhinopneumonitis (herpes). Other viruses and bacteria may also be involved, however, and in many cases infection is due to more than one cause at a time. Other important causes of coughing which are not infectious include pharyngeal lymphoid hyperplasia and COPD.

Diarrhoea: Worm damage to the colon and caecum, salmonella infection, other bacterial infections, liver disease, poisons and tumours are all possible causes. Treatment with antibiotics may also produce diarrhoea. Too much fresh grass may produce a less severe type of diarrhoea but is a common condition.

Foals often have mild or transitory diarrhoea which responds quickly to minor treatment. Specific causes include 'foal heat scour' which occurs when the mother comes into her foal heat, seven to ten days after foaling, bad feeding, bacterial or viral infections (some of which may be of a more serious nature eg salmonella) or worms (Strongyloides westeri). In more serious cases the affected foal becomes dull and lifeless, develops fever, rapid pulse and respiration and will not feed. Untreated serious diarrhoea will lead rapidly to dehydration, collapse and death. Vital factors in maintaining young foals in good health are good stable hygiene, and ensuring an adequate supply of colostrum, the special milk produced by the mare in the first day or two after foaling which contains antibodies and is therefore responsible for transferring immunity to the foal.

Treatment of diarrhoea includes giving fluids, either by mouth or intravenously, kaolin and other absorbents, steroids, antibiotics and wormers. It may also be useful to give treatment aimed at re-establishing the normal bacterial population of the bowel, and this may be done by giving live yogurt or bacterial preparations which your veterinary surgeon may use.

Equine Herpes Virus Infection (Also known as EHVI or Rhinopneumonitis): This virus can cause respiratory disease, abortion, early death of foals, or nerve damage. The form of the disease causing nervous disorder may follow respiratory symptoms or abortion, or it may occur at the same time. Affected horses show a weakness of their hind legs, paralysis of the tail, incontinence, and progressive inability to stand. Many cases will not respond to treatment.

Equine flu: See the section on vaccinations for details of this disease (p28).

Exertional myopathy ('Tying-up syndrome', 'set-fast', azoturia): This used to be seen typically when carthorses fed on a high carbohydrate diet over the weekend returned to work on Monday. It is also seen these days in horses which are on good diets but which are then over-exerted. Horses involved in endurance rides may also be affected.

Signs of tying-up are a shortening of the horse's stride and stiffness; more severe cases may lose their appetite,

sweat and show pain. Horses out at grass sometimes show stiffness and reluctance to move but initially maintain a good appetite and show no signs of pain. However, as the condition progresses the horse becomes unable to move and death may follow. The urine becomes dark brown or blood-coloured due to muscle breakdown products (myoglobin) being filtered through the kidneys.

The signs of this condition might be confused with laminitis, colic or tetanus – accurate diagnosis is usually made from blood and urine samples. Treatment includes fluid therapy and anti-inflammatory drugs as well as changes to the diet.

Grass sickness (Dysautonomia): Many forms of this disease are found, varying from emaciation and weakness to sudden colicky signs and rapid death. The cause is unknown, but it is more common in certain areas and in horses out at grass in the summer months. Diagnosis is based on the clinical history and local occurrence of the disease as well as the clinical signs. However, confirmation is often only possible at post-mortem examination and there is no specific treatment. The majority of affected horses will die from it.

Similar signs are seen in cases of colic or choke.

Guttural pouch infection: The guttural pouch is a sac originating from the horse's eustachian tube (which runs between the inner ear and the pharynx). Infection of the pouch may be bacterial, causing a discharge of pus from the nostrils when the horse lowers the head; or fungal, which often damages the nearby blood vessels and in some cases leads to fatal bleeding from the nose. Diagnosis is usually made by passing an endoscope into the pouch via the nose. Treatment involves surgical drainage of the pouch and irrigation with appropriate drugs, but is often unsuccessful in the case of fungal infection.

Harvest mite (Trombicula): The mites can be seen if you use a good hand lens on the skin around the heels in the late summer or autumn. They cause inflammation and irritation, but are easily dealt with by a variety of drugs which are usually used in the form of washes or creams.

Lameness: The diagnosis and treatment of lameness is an enormous subject which cannot be covered here. Attempts to diagnose lameness yourself are dangerous because an incorrect diagnosis may lead to danger to the rider or more serious injury to the horse if the real cause remains undiscovered. It is also important to be able to rule out nerve disorders and other non-orthopaedic conditions.

It is probably best for the horse owner to stick to a thorough inspection of the foot and limbs for any visible abnormality, and check for such things as flints lodged in the foot, minor wounds or loose shoes before calling the vet. Some specific causes of lameness are mentioned elsewhere in this section.

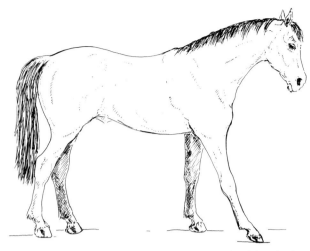

If a horse 'points' like this it is a sign of pain in the front foot or leg of the pointed limb. Sometimes the horse will rest the foot on the heel and occasionally will put the leg slightly backwards and rest the weight on the front of the foot at the toe.

Laminitis: The old name for laminitis was 'founder'. The disease involves inflammation of the sensitive laminae of the hoof, the tissue between the bone of the foot and the hoof wall. It is particularly common in overweight ponies kept on good pasture and not given enough exercise; it is also seen in horses fed on too much grain and in mares following a uterine infection. The blood supply to the foot is affected following the initial inflammatory stage and the laminae become damaged, with defective production of the horn of the hoof and rotation of the pedal bone also occurring.

In the initial stages the foot is hot and painful (often the fore feet are more affected) and the horse sits back on his heels to take the weight off the front feet. Other signs of pain such as sweating may also be seen.

Long-term damage due to poor horn production and rotation of the pedal bone may continue to make the horse lame. At this stage x-rays are useful to assess the degree of rotation of the bone and associated damage. Correct trimming of the hoof, to lower the heels and bring the pedal

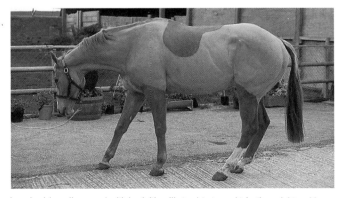

An animal (usually a pony) with laminitis will stand to try and take the weight on his heels, away from the most painful area of the foot at the front. This pony is affected in all four feet, to show the stance, but normally only one pair of feet is affected and usually the front, although sometimes only the hind feet are involved.

bone more nearly parallel to the ground, and surgical shoeing are important forms of treatment at this stage.

The horse or pony should be put onto a very much reduced diet. Hosing or bathing the feet may be useful initially. Walking the horse is not helpful once hoof separation or pedal bone rotation have begun. Anti-inflammatory drugs, painkillers and fluid therapy may be given by your vet.

Lice: Lice and their eggs (nits) can be seen easily on the hair. They cause irritation and rubbing, with loss of hair and thickening of the skin, but are easily treated with a variety of products.

Lymphangitis: Following a wound or other infection, or as a result of other damage to a horse's leg, the tissue may become very swollen and puffy; in severe cases a yellowish, sticky exudate may ooze through the skin. It is this latter condition which is known as lymphangitis; diagnosis and treatment are more concerned with the primary condition.

Lymphoid pharyngeal hyperplasia: This is a condition seen mainly in young horses. It affects the reactive tissue at the back of the mouth and can be a cause of coughing. Diagnosis is based on the appearance of the tissue seen through an endoscope. There is no specific treatment and it will usually improve with age.

Melanomas: are a type of tumour commonly found under the tail of older grey horses. The growths have a smooth, darkly pigmented surface with a domed shape, and grow fairly slowly but will spread in time. If there are multiple tumours surgery is of limited value.

Mud fever (Cracked heels): This is a bacterial infection common in wet or muddy conditions. The skin becomes chapped, and infection can then get into the skin causing a severe dermatitis with scabs forming on the surface. The leg becomes sore and swollen and the horse may become lame. Treatment with antibiotic creams and thorough cleaning of the skin is usually effective. The same bacteria may also cause 'rain scald' on the horse's back.

Navicular disease: is usually found in horses aged from about six to twelve years, especially if worked irregularly and on hard ground. The disease process involves damage to the blood supply of the navicular bone (which lies just behind the joint between the pedal bone and the second phalanx). Usually the lameness develops over a period of time and affects both front feet.

Initial investigation may involve using nerve blocks to anaesthetise the back of the foot – if the pain is in this area the horse will then go sound. However, it is usually essential to x-ray the foot to make an accurate diagnosis of navicular disease and to differentiate it from pedalostitis. Once the diagnosis is confirmed, treatment is aimed at improving the bone's blood supply as well as controlling pain.

Several different approaches to treatment have been used. Many cases were successfully treated with warfarin, an anticoagulant which therefore prevented blood clotting and blocking the blood vessels to the bone; but this has been largely superseded by drugs which dilate the blood vessels, principally isoxuprine. This second approach to treatment is far less risky for the horse than warfarin (which is also used in rat poison), and treatment therefore needs less intensive supervision. However, many cases of navicular disease will never remain consistently sound.

Pedal ostitis: In pedal ostitis the bone of the foot becomes demineralised in response to inflammation. Persistent bruising of the sole or laminitis may lead to pedal ostitis, though there are other possible causes – some cases may be due to infection. Treatment is aimed at dealing with any specific cause and protecting the sole from bruising, as well as dealing with the inflammation. However, many affected horses will continue to deteriorate.

Periodic ophthalmia (Moon blindness): It is not certain what causes periodic ophthalmia. It is uncommon in horses under four years old. It affects the iris and associated structures of both eyes, and has a very marked tendency to re-occur. The horse will turn away from bright light and hold the eyelids closed, and there may be a profuse and watery tear discharge; other specific signs can be detected by your vet. The condition may lead to more severe damage such as cataract. It is sometimes difficult initially to differentiate periodic ophthalmia from severe conjunctivitis or wounds of the eye.

Photosensitisation: can occur as a result of eating specific plants eg St John's Wort, or as the result of liver damage. In either case an accumulation of chemicals in certain areas of the skin causes damage when these are exposed to sunlight. It affects only unpigmented areas of skin, especially a blaze and the lower limbs, and causes a severe dermatitis. The main element of treatment is to stay out of the sun, though sun-screen creams or tattooing are sometimes effective.

Ragwort

Ragwort poisoning: Most horses will not eat fresh ragwort, but they may try it if it has been left lying after being pulled up or is found in hay. It causes severe liver damage which develops slowly, though signs of the condition usually appear quite suddenly: the horse seems to become unaware of its surroundings, it wanders around in circles, becomes aggressive or appears blind. Often there is jaundice, weight loss and diarrhoea.

There is no curative treatment and most horses will die once signs have developed. Ragwort should be pulled up and *burnt*, and pasture should be chemically treated to eradicate the plant.

Ringbone: Arthritis around the pastern or coffin joint is known as high or low ringbone respectively. It is usually due to injury to the membrane around the bone (the periosteum) as a result of ligament injuries or direct damage. It is more common in the front feet and the signs of lameness are not specific. The joint may be warm or swollen, and movement of the joint may be limited. Nerve blocks may help to show the area affected, but x-rays will give the most information about the condition. Treatment with pain-killing drugs may keep the horse working, although some cases may not respond at all – then surgery to fuse the joint solid may be successful; however, some cases may have to be destroyed.

In the early stages of the disease some horses will benefit from being shod with shoes with a rolled toe. In a few cases the joint may become ankylosed (fused) naturally and the lameness will improve, though since the movement of the joint is of course abolished the lameness may continue to some extent.

Ringworm: is a fungus, not a worm. It causes hair loss and also crusty circular patches of skin to develop anywhere on the horse but most commonly around areas in contact with the tack. The fungus survives very well on leather, fence posts or stable woodwork and can be very difficult to get rid of. It is highly infectious and tends to spread quickly through a stables; it is also infectious to people. A number of different drugs are available for treatment, and it is also very important to treat the tack and any other reservoir of infection.

Roaring: Roaring is a noise which can be heard in horses with laryngeal paralysis, mainly when they are worked hard. The paralysis of the larynx is due to damage to the left recurrent laryngeal nerve and this allows the cartilages of the larynx to collapse inwards, partially blocking the airway. The roaring noise is produced by the vibration of the cartilages as the horse breathes in. Diagnosis can be confirmed by observation of the larynx using an endoscope. Treatment is usually surgical. The operation is called ventriculectomy or Hobday's operation and involves removing the laryngeal saccules which lie between the cartilages in order to draw the cartilages back to a more normal position.

Spavin: Two types of spavin are important, bone spavin and bog spavin: bone spavin is an osteoarthritis of the hock joint. It causes pain and lameness, but may be treated in a variety of ways with analgesic drugs. A spavin test is quite useful: the leg is held up with the hock flexed for about two minutes, after which the horse is sent off sharply at a trot – an affected horse will show an exaggerated degree of lameness. X-rays of the joint will give more precise information about the extent of the problem. In many cases the horse starts off lame and improves as he warms up. Treatment will often allow a horse to go on working for some time after spavin first appears. Blistering, firing or neurectomy (desensitisation by cutting the nerves) are treatments which are entirely out of date, although you may see them mentioned occasionally in older books.

Bog spavin is a distension of the joint space itself without bony involvement. It often does not seem to cause lameness except in severe or very recent cases. It may be a sign of poor conformation, injury to the joint or nutritional deficiencies. Many yearlings show a transient bog spavin which disappears as they get older. In the majority of cases no treatment is necessary.

Splints: The splint bones lie against the back of the cannon bone and are joined to it by a ligament. Inflammation resulting from injury to the ligament and the membrane covering the cannon bone causes a bony reaction on the splint bones. While developing, these 'splints' are associated with lameness but in most cases the soreness and lameness disappears in time, although the lumpy swelling can still be felt. They are most common in young horses. The splint bones may also be fractured, perhaps by a kick or similar injury.

Strangles: usually affects horses under about five years old, and is a bacterial infection which is quite contagious. The bacteria which cause it are very resistant outside the animal and may survive in a stable for up to a year. Signs include a fever, cough, nasal discharge, lack of appetite, and discharge from the eyes. The glands (lymph nodes) under the angle of the jaw become swollen and develop into abscesses which may rupture to discharge pus. In a few cases there may be internal spread of the disease with abscess formation. Treatment with antibiotics combined with surgical treatment of abscesses is usually successful.

Sweet itch: is an allergy to the bites of certain types of midges and occurs in the summer months, especially in ponies. There is usually intense irritaion along the mane and the base of the tail and the horse may rub itself quite raw. It is advisable to keep susceptible ponies away from ponds and marshes and to maintain good fly control in the stables.

Tendon injuries: Tendon strains are relatively common injuries and vary from mild over-extension damage or bruising to complete disruption of the tendon. Mild

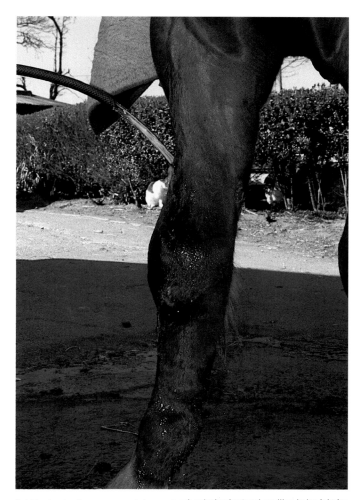

Cold hosing is often recommended as a way of reducing heat and swelling in leg injuries.

Thrush, a foot infection caused by wet and/or dirty conditions or badly kept feet, needs attention from vet or farrier.

injuries will be helped in the initial stages by cold applications to limit the inflammatory reaction. Later, once the initial inflammation has subsided, warm poultices or liniments or embrocations may help because they stimulate a mild inflammatory response which encourages the healing process to continue. In more severe cases specific treatment with drugs may also be needed. In general the most important treatment for mild tendon injuries is rest, to allow the tendon to heal fully before subjecting it to further stress. A principal aim of treatment is to allow repair of the tendon by healthy tissue rather than by scar tissue, which is weaker than the original tendon. In more serious cases surgery may be necessary to repair disrupted tendons or to reinforce them with man-made materials, such as carbon-fibre. Following initial treatment it may be necessary to put the leg in plaster for support while healing takes place. Various forms of physiotherapy such as ultrasound or laser treatment also seem to have a worthwhile place in the rehabilitation of these injuries. However, majority opinion holds that firing and blistering have no place in modern treatment and most veterinary surgeons would probably consider that such practices cause quite unnecessary suffering to the horse.

Tetanus: (See under vaccinations (p28).

Thoroughpin: is a swelling in the hock joint which appears just above the point of the hock and in front of the Achilles tendon; when the swelling is pressed it will move to the other side of the hock. The swelling is a distension of the sheath of the deep digital flexor tendon, and in many cases there will be no lameness. Treatment is often unnecessary but in a few cases the swelling needs to be drained, possibly with injections of a steroid into the sheath at the same time.

Occasionally when due to injury there may be other associated damage which needs attention.

Urticaria: Many different substances may produce urticaria, which is an allergic reaction. It affects the skin which typically shows a 'nettle-rash', with raised areas of varying sizes. It will often disappear spontaneously, but some cases may be severe and show more general illness which requires veterinary attention.

Windgalls: Windgalls are fluid-filled distensions of joint capsules or tendon sheaths, and the term is usually applied to swellings around the fetlock joint. They are not normally associated with lameness and are common in horses which work hard. In most cases no treatment is necessary.

Wobbler disease: Wobbler syndrome is a colloquial term covering a variety of conditions affecting the spinal cord in the neck. The most common of these is due to compression of the spinal cord as a result of narrowing of the spinal canal. Affected horses become incoordinated with their hind feet and may be partially paralysed. An affected horse is not safe to ride.

NURSING

Sick horses need to be kept dry, clean and warm and offered a variety of palatable food and clean water to drink. A sick horse should be kept on his own in a good-sized loose box with a deep bed, there should be some means of providing extra warmth if needed, and ventilation should be good but without draughts. The box should be kept scrupulously clean. Make sure that any heating device does not present any fire risk.

It may be necessary to use rugs and blankets to keep the horse warm, in which case several light rugs are better than one heavy one. Rugs worn overnight should be changed in the morning, and all rugs should be kept clean and dry.

Make sure that all food offered to the horse is fresh and good quality; try a variety of different foods and give small quantities at a time. Make sure that the water buckets or trough are kept filled and clean, and change them often so that the water is always fresh; take the chill off the water before giving it. Take away any left-over food, and scrub buckets or mangers after each feed to avoid any stale smells or debris. Ask your vet about any special feeding needs, but in general stick to top quality feed and fresh greenery such as carrot tops or freshly picked grass.

Gentle grooming with a soft brush should be done daily, and it is essential to bathe away any discharges from the nose, mouth, eyes or anus.

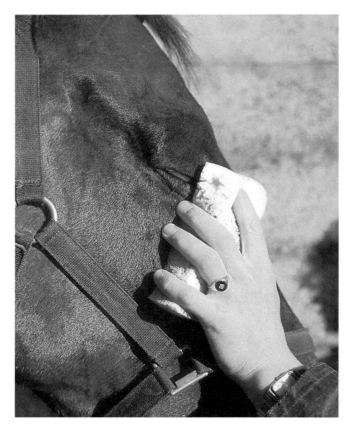

Remember that the sponging part of grooming is very refreshing.

FIRST AID

It is very important that all those caring for horses know how to administer correct first aid to sick or injured horses and how to check their general health. Spotting sickness early can save an animal's life by enabling treatment to be started promptly. Similarly, correct early treatment of injuries hastens efficient healing and, in more serious cases, aids subsequent veterinary attention. Never hesitate to call in expert help if in doubt.

Colic

Remove food from the stable or transfer the horse to a bare paddock. If the horse is stabled make sure that the bed is deep and bank it up well around the edges of the box so that if the horse rolls he will not get stuck, or 'cast'. It may be helpful to walk the horse in order to prevent rolling but otherwise it is not necessary. Make sure that he is kept out of draughts if he has sweated up and use rugs if necessary. If the horse's respiratory rate or pulse is increased or if he is sweating you should call your vet at once. Any colic which has not improved within two or three hours should be seen by your vet unless already instructed otherwise.

Electrocution

Always look out for bare wires or broken electrical apparatus, especially in older stables. Make sure that the power is turned off before handling the horse. Keep the horse warm and call the vet.

Eye injuries

Gently try to part the eyelids and look for any sign of a wound or foreign material such as a hay seed or thorn. However, do not try to pull a foreign body out as it may be barbed or hooked, which will cause further damage. The best method is to try to flush the eye with a weak saline solution to dislodge the foreign body. Even when this is successful you should call your vet to check for any associated injuries and to treat the infection which will almost certainly have been introduced.

Haemorrhage

Horses are large animals and wounds may produce a great deal of blood. Also a little blood goes a long way! The first essential in dealing with a haemorrhage is not to panic. Call assistance, and use pressure to control the bleeding, preferably by firmly applying to the wound a pad of gauze or similar material soaked in cold water and squeezed dry. Imagine that you are trying to flatten a squash ball with your hand in order to gauge the pressure needed. Leave the pad applied for at least two minutes and resist the temptation to lift it off to see what is happening. Timing yourself with a watch is very helpful. Tourniquets are not advisable as they are extremely difficult to apply properly.

Heat stroke

This may occur at any time, but especially on endurance rides or during polo games. The horse should be cooled by using large quantities of cool water, particularly over the head and neck. The water should be wiped off again frequently and more applied, preferably by a hose, as the water nearest the skin will otherwise provide an insulating layer.

Lameness

Before calling the vet check for loose shoes, stones wedged under the shoe, flints embedded in the sole or nail wounds. If after resting the horse a minor lameness persists for more than two or three days, you should call your vet or you may find that a minor problem turns into a major one. Any lameness associated with obvious pain, or any severe lameness should be seen as soon as possible.

THE FIRST-AID KIT

What should you keep in a first-aid kit? The following list is probably sufficient for most purposes:

1. *Some pads of gauze or lint* which can be used soaked in cold water as cold compresses to control bleeding, or dry as dressings to be used under a bandage. They should be about six inches square and should be kept in sealed packets ready for use.
2. *Bandages* are useful, to provide a compression bandage or to hold a dressing in position. They should be long (about eight feet) and four to five inches wide, slightly elastic (more so than the usual sort of stable bandages), and absorbent. Keep them rolled with the tapes on the inside ready for use.
3. *Cotton wool:* rolls of 'hospital quality' cotton wool are most useful. It can be used for cleaning wounds and bathing away discharges and also to provide padding under a compression bandage.
4. *Salt* for making up saline solution. For cleaning wounds one teaspoon of salt to a pint of warm water is about right. Your vet may want you to use a stronger solution of one teaspoon to a cup of warm water for bathing swellings or contaminated wounds.
5. *Disinfectant* such as cetrimide (Savlon) which can be used diluted for bathing wounds, or more concentrated for disinfection of buckets, utensils or tack.
6. *Wound powder, spray or ointment.* The ubiquitous purple spray found in most stables is not necessarily the best thing to use for some wounds, and in many cases powders and ointments are potentially harmful. Most such products are prescription drugs and should only be used on veterinary advice.
7. *Poultices* for dressing minor wounds or older tendon injuries.
8. *Fly repellent powder or spray* is very useful in the summer months to prevent contamination of minor wounds.
9. *Wormer.* Not really for first aid but if you keep a dose in your medicine box you won't forget about it.
10. *A pair of pliers or wire cutters* can be very useful in an emergency.

Never 'clean' a wound with a sponge like this as you will almost certainly introduce more germs. Always use a fresh piece of cotton wool or Gamgee Tissue or bathe by gently syringing saline solution or slightly medicated water on to the wound. Suitable syringes can be bought from your vet or a veterinary chemist: made of plastic they are quite inexpensive and very useful.

ROAD ACCIDENTS

The first priority is to control the injured horse and direct traffic. A head collar, halter or bridle should be put on and the horse held on a lead rein, preferably away from the road. In some cases where there is obvious severe injury, such as a broken leg with other injuries to other limbs, it may be better to keep a fallen horse lying down by restraining him with weight across his neck, either by someone lying across the neck or by using sacks. It is still important to use a halter or bridle in case the horse manages to get to his feet, otherwise control will have been lost. Obviously you should call the vet at once. The police should also be called if a vehicle is involved.

Respiratory distress
If your horse is having difficulty breathing make sure that the nostrils are clear of discharge and that the stable is well ventilated — take care to avoid draughts, however. Increasing the humidity of the air may be helpful by using steam, though it is important that anything used to generate steam cannot be kicked over or cause injury. It may be useful to add an inhalant such as turpentine oil to help relieve congestion. Inhalants containing menthol or coal tar derivatives may not be liked by some horses but can be tried. Any case of respiratory distress is potentially serious and veterinary attention should be prompt.

Tendon injuries
If you find a hot, painful tendon injury of recent origin it is helpful to hose it with cold water and to apply a compression bandage to give support. It is also important to bandage the opposite leg to give additional support, and keep the horse confined to a stable. If pain or swelling persist for more than twenty-four hours you should call your vet.

Wounds
The first priority with a wound which is not bleeding is to clean it as soon as possible in order to reduce bacterial contamination. Saline solution or weak disinfectant such as Savlon are suitable agents to use. Alternatively, hosing a wound is a useful method of flushing out debris — though take care not to drive foreign material deeper into the wound.

Minor wounds can be treated by poulticing. Such wounds may not need veterinary attention but it is important to bathe them at least daily and prevent scabs forming. If any swelling or discharge develops you should call your vet.

There are four ways of keeping horses: stabled; at grass; on what is called the combined system, where the horse is partly stabled and partly at grass; and yarding, where horses live in surfaced enclosures with no grass.

The horse evolved as a creature of the wide open spaces, a natural nomad on the move much of the time, so it is understandable to think that the best system for him is to be at grass all the time. In wild conditions, and if the type of horse or pony is indigenous to the environment in which he finds himself, this could well be true. But domesticated conditions are normally much more restricted, sometimes even cramped, with fields often overstocked and carrying many more animals than the grass growth can support. Usually there is nothing like enough shelter, either natural or man-made, and the ground can become bog-like and badly poached (churned up by too many hooves) in wet weather.

The other end of the spectrum is to keep a horse stabled, but this is obviously completely unnatural to the horse and can be extremely stressful to him, particularly if he is kept short of exercise; and most privately owned animals *are* short of exercise because their owners have other commitments, or maybe because they fail to understand just how essential movement and space are to horses.

Very many people feel that the best system for both horse and owner is the combined system: the horse is stabled when necessary or desirable, but is also turned out for freedom and exercise, and for natural contact with other horses and ponies, provided these are turned out too.

There is another system which can be excellent when there is neither the labour nor the desire to stable horses but where grazing is at a premium, and when for some reason (such as work) the horses need to be kept handy: this is called 'yarding' in Britain and Ireland, the 'open stable' system in Germany, and 'coralling' in America and Australia. It is not very common in the UK, is more so however in Ireland and continental Europe, and very common in the western states of the United States of America and in the farming and ranching areas of Australasia.

Let's look at the advantages and disadvantages of these methods.

STABLING

In this system the horse is a prisoner, totally dependent on humans for his every need, and since horses are labour intensive this is also the most time-consuming and expensive method of keeping them. Kept like this the horse is completely dependent on us for his food, water, bedding, grooming, clothing, exercise and general health, and it is our responsibility to see meticulously to those needs and to acquire the fairly high level of knowledge needed to understand what to do for the horse, how and when, and to understand when specialist help – such as from a vet or farrier – is needed.

A healthy, fairly fit horse needs a good two hours exercise a day, although many do not receive so much. However, this amount is *much* less than the sixteen hours a day or so a horse at liberty would spend exercising himself, so you can see just how restrictive stabling is. There are some horses who simply cannot stand being cooped up in this way and show their distress by psychological reactions as well as physical ones. At the very least they become bad tempered, or develop neurotic behaviour vices: these vices include crib-biting (see p22); wind-sucking where the crib-biter has perfected his art to such an extent that he can gulp air down without having to grasp a solid object first; box-walking, where the horse walks round and round and round his box almost as if in a daze; and weaving, where he stands usually with his head over the door, but sometimes inside the stable, and swings his weight from side to side from one foreleg to another and his head with it. Fortunately those horses who are so badly affected by stabling seem to be in a minority; however, many more show their anxiety by becoming tensed up and generally unhappy.

Amazingly, bearing in mind the wandering nature of the animal, very many horses *are* content with this management method, particularly if they are well exercised and have a stable in a position to suit them, either busy or quiet as the case may be; usually these horses have reasonably steady personalities and are not highly-strung – and this does not necessarily depend on breed. There are many quiet, placid Thoroughbreds who adapt well to stabling, and there are nervy natives who do not.

The time devoted to ridden exercise can be halved by experienced riders by riding one horse and leading another, and the horse can also be exercised by lungeing, loose schooling, or long-reining. In lungeing the horse is attached to a long rein or line (usually webbing) held by his groom or trainer and made to go round and round the trainer in a circle, at various paces commanded by him or her. In loose schooling the horse is free but works to command either in an indoor school, a well-fenced outdoor enclosure, or perhaps down a high-fenced 'lane' or corridor – this last method is usually used for exercising or training jumpers.

Another time-consuming daily chore is mucking out the stable. Droppings and urine-soaked bedding have to be removed at least twice a day, and to muck out an average loose box properly takes about half-an-hour – if you're used to it and fairly fit.

Grooming a stabled horse has to be done thoroughly to keep him clean and to maintain his skin in the tip-top condition needed by a working horse, and this can take at least half-an-hour a day and usually an hour if you include the brush-overs necessary at other times during the day.

In addition to these most time-consuming of tasks, the horse has to have his feeds mixed, his hay nets or rack filled, his water buckets cleaned out and refilled several times a day or – though less often – his automatic waterer attended to, and his feed buckets and manger cleaned daily to prevent caked-on, old food harbouring disease. His clothing, if worn, needs brushing, shaking out, folding and putting on and off, and laundering fairly regularly, and turn-out rugs for wear in winter in the field can take a considerable amount of time to clean and dry.

His saddle and bridle will need cleaning (a good half-hour of a job), the tack room and feed rooms should be kept reasonably tidy, the yard swept and the walls and ceiling of the stable periodically swept or vacuumed out.

MECHANICAL HORSE WALKERS

Mechanical horse-walkers are increasing in popularity. Some consist of overhead arms to which the horse's lead rope is fastened; others of compartments in which the horses are free, and can be commanded by one supervisor to walk, trot or turn as desired. In one model the horses follow a track around the perimeter of a manège or indoor school, which does make a change from going round in constant circles. The circular models can take up to eight horses, although the most popular take four or six; the other type can take more.

Only in the horse-walker method is human labour significantly reduced, and even then the time advised (up to half-an-hour) is only part of a horse's daily requirement.

What you might call a natural jumper! Not all horses will go loose into a manège like this and make a bee line for any jumps available of their own free will, like Rags, but most will do so, or at least enjoy the experience, once taught what to do. It is a valuable and enjoyable form of exercise which makes a change from being under constraint when working.

The advantages of stabling to the horse are warmth and ⸻ bad weather; a cool escape from heat and flies in summer, though this depends on ventilation; something resilient and dry to lie on and, if he is well cared for, a reliable supply of food and water. He can often see and maybe talk to other horses also in the yard, and it is true that many horses come to regard their stables as little havens where there is food, water and shelter, not to mention freedom from field bullies.

The advantages to us are several: the horse is always readily available and reasonably clean when we need him, and is easier to *keep* clean, too; we can control his diet minutely should we feel this necessary; and by clipping off his winter coat and by exercise we can bring him to whatever stage of fitness we need for the work we require of him.

KEEPING A HORSE AT GRASS

The horse still cannot be left entirely to his own devices, but this method *is* far less time-consuming and tying for the owner. Every day, the horse will need checking over for illness or injury; the field boundaries (hedges, fences and gates) will need checking for breakages and gaps through which the horses will be only too thrilled to escape and go walkabout in neighbouring crops or gardens; the water supply will need checking, and ideally droppings should be removed from the land daily, although this may be a counsel of perfection. If there is a field shelter – which I believe there must be for horses out for more than a few hours a day, particularly at times of extreme weather in either winter or summer – then droppings must be removed from this every day, too; and a weather eye must always be kept for litter which so often mysteriously arrives in the field: beer cans, bottles, polythene bags, old fertiliser sacks (which can make horses very ill, even if they only lick the residues), newspapers blowing about which can panic horses into crashing through the fence, and implements such as shovels, farm machinery and so on which may be left in the field. If the field borders residential gardens, poisonous garden waste may be dumped in your field, and there are many garden plants which are very poisonous to horses.

This all sounds a lot, but it takes up very much less time than attending to a stabled horse. If you are riding your grass-kept horse, the exercise time (which is supposed to be enjoyable!) must be added to the list, plus time to lightly groom the horse and, of course, clean tack and turn-out rugs if used.

There are things on the plus side: at certain times of year – such as winter, or in summer during a drought which affects grass growth – you'll need to give regular feeds, especially of hay, to the grass-kept horse, but normally you are not *so* tied to a particular time, as there will usually be at least a bit of something to keep the horses busy and stave off their hunger. Regularity, however, is still to be aimed at, as horses do come to know and expect a routine.

With this method there is no actual compulsion on you to ride if you don't want to, as the horse can exercise himself, although he won't keep fully fit in the field. However, two or more horses out together, perhaps corn-fed depending on the time of year, will keep each other half-fit because their social intercourse, games and so on keep them on the move; but a solitary horse may well just stand by the gate feeling lonely and miserable, just wanting to be brought in where he'll find company and things to keep him interested.

Given reasonable care and conditions, grass-kept horses usually stay healthier, if not fitter, than stabled ones, due to the constant gentle exercise and fresh air – disabling respiratory problems are particularly common in stabled horses because the ventilation in most stabling is normally so poor.

From the horse's point of view the main disadvantage with this method is the frequent lack of really effective shelter. Horses suffer readily from exposure in fenced, domestic paddocks where there is neither windbreak nor overhead shelter.

Over-confining horses by leaving them stabled too long without proper exercise and, particularly, freedom, causes tension and distress, one sign of which is damage to stables of the sort shown here where the horse has scraped the paint off the door with its teeth and, when the door has been closed, bitten the wood inside the top door.

Ideal wooden post-and-rail fencing and even, well-kept pasture. Note how the rails are run on the *inside* of the posts to protect the horses from the latter. Similarly with the protective fencing round the young trees, planted for future shelter: the fencing protects the trees from the horses stripping their bark which can kill them, but the rails again protect the horses from the fence posts.

Removing droppings as often as possible from paddocks, especially small ones, is an important part of stable hygiene. Here the droppings scoop and shavings rake shown on p56 are being used.

This is *not* a natural situation for them, as in the wild they are free to roam over many miles, seeking out sheltered places when they feel the need. Moreover, the poached ground so common in our British winters causes mud fever and cracked heels, both potentially serious conditions which can cause blood poisoning if neglected, not always easy to clear up, and a cause of lameness and great discomfort and soreness to the horse.

Should the field be over-grazed, badly managed and not rested, usually due to overstocking with too many horses, the grass supply can dwindle to next to nothing even though to our eyes there may still seem to be edible grass growing; then the horses obviously suffer from hunger, but since they are fenced in, they do not have the opportunity to wander off to pastures new to appease their hunger. This is when some may try to push their way out in an effort to find something to eat – though many more just lose condition, becoming run-down and susceptible to disease, not to mention miserable.

Horses have a well-defined herd hierarchy, with leaders, followers, whipping boys, bullies and every other category all of which can be witnessed in any school playground. If a horse is being bullied, either passively (when the others refuse him access to food, shelter, water and company) or actively (by kicking, biting and so on) there is no route of escape as the bullies simply follow him round the field making his life a misery.

Another serious disadvantage is being pestered by flies in summer. Very many people seriously underestimate or simply disregard the sheer torture of horses plagued by flies and other insects; yet without proper shelter they have no escape, unless their owner is compassionate enough to use a really effective, long-acting fly repellent on them. Flies bite and irritate horses until in desperation they stampede – their only relief is to gallop faster than the flies can fly. But when they stop the flies land again, feeding on their sweat and the discharges from their eyes in particular, and attracted to all the thinner-skinned, sensitive areas of their bodies such as genital organs. The horse's natural defences against them are pitifully inadequate and extremely temporary: a shake of the head, flick of the mane or tail, a stamp or a twitch of a muscle may dislodge a fly, but it will land again instantly. Their eyes can become very inflamed, sore and may even ulcerate;

their legs may become seriously jarred and can cause lameness, and their feet become cracked and sore through stamping or galloping on hard ground – and of course they suffer from exhaustion and terror through constant galloping and the impossibility of escaping their pests.

Some flies suck blood and cause inflamed weals which itch, are sometimes rubbed raw and may become infected; others lay eggs on the horse's coat which it then licks off and swallows: they hatch out in the stomach where the larvae cause damage and colic (indigestion). One particularly distressing condition, sweet itch, is caused by an allergy to the saliva of a midge: the horse itches and rubs itself on fences or trees, again until it is raw.

On the plus side (and you were probably wondering if there really was one!), the horses are not short of exercise and – hopefully – company, and usually have a good supply of grass, their natural food, and also water. If the field has ample natural shelter in the form of thick, high hedges on the windward side of the field, also overhead shelter in the form of several mature trees which form a dense overhead canopy, and particularly if the horses are in fields of ample size (preferably ten acres or more) they may certainly find they are living in the closest possible domesticated version of Trapalanda, the horses' heaven. Horses become very independent of their owners under such conditions, and some people find it quite hurtful when they go to check on their horse, bearing gifts, only to be ignored or treated with no more than a passing interest! However, once conditions deteriorate, back comes the horse cap in hand, wanting all the attention and creature comforts!

COMBINED SYSTEM

Under normal circumstances this system is one of the best for both horses and owners. Usually the horse is out for part of his twenty-four hours, and in for the remaining part; a good method is to leave him in during the summer days and out at night, reversing the process in winter so he can have winter nights in the shelter of his stable. However, the actual timing of going out and coming in is fully flexible according to the circumstances of the owner and the needs of the horse.

For instance, a working owner may be unable to ride on winter days but would rather not subject her horse to a long, maybe wet and cold day in the field, particularly if there is no shelter; in this case it may be possible to turn him out in the morning, and arrange for someone else to bring him in a couple of hours later to a waiting haynet, then turn him out again in the afternoon for an hour or so, for further essential exercise. In summer, on the other hand, the horse may need to

Adequate turn-out facilities are important to a horse's health and mental contentment and correct fencing goes a long way towards ensuring, as far as is reasonably possible, their safety when out.

This is an example of correct post and rail fencing. The rails are level with the very tops of the posts, which slope down to facilitate rain run-off so lengthening the life of the posts. The top rail is as high as the horses' backs, the bottom one high enough to prevent their getting legs over and injuring themselves, yet low enough to stop a foal rolling under and getting up on the other side in a panic.

Rails are always best on the insides of posts to stop horses pushing them off the posts by leaning on them, but here, where one line of fencing separates two paddocks, two lots of rails may be prohibitively expensive. However, to lessen the chance of horses injuring themselves on the exposed posts when galloping along the fence line, as horses do, a rail should be run along as shown here at about the height of a horse's point of shoulder. As an added precaution the gap between this rail and the fencing should be filled in to prevent a horse getting a leg caught between the two.

CARE AND MANAGEMENT

Yarding
Accommodation and facilities
Stable bye-laws

Horses suffer greatly from the attacks of flies in summer. They always attack worst the most sensitive areas and particularly the eyes, feeding on the discharges. This creates inflamed eyes which run even more and, in some cases, can become infected and ulcerated. The vicious circle increases as the flies come in larger and larger numbers to feed on the increasing secretions. It is most important to use an effective fly repellant which is, however, safe to use around sensitive areas and your veterinary surgeon should be able to help you find one. You can also contact the manufacturers of fly repellants you see for sale to check their suitability.

come in only a couple of hours before being ridden, to allow the grass he will have been eating to digest a little before work. It's entirely up to you how you decide to do things.

Horses are quite adaptable, and although they like a reasonable overall routine they do sense that it varies on different days. For example, they quickly catch on to the fact that there is a different routine at weekends or on show days; but provided their meals are fairly regular and their diet is not suddenly changed (which seriously upsets their digestive systems), they can easily adapt to somewhat erratic hours in and out of the field. One precaution which *is* wise, however, is not to keep your horse off grass for several days and then suddenly give him a full day in the field. He should really have some time at grass every day, or at least be given half an hour's grazing in hand on the days when he is not going to be turned out, perhaps because he is hunting or at some event. This way there will always be some grass going through his system, and you won't commit one of the gravest sins in horse management – to change his diet suddenly.

The combined system means the horse can be kept as fit as a stabled horse with a controlled diet (you can, after all, limit the amount of grass he gets), and virtually as clean. A clipped, fit horse can be turned out with a good turn-out rug and maybe a neck and head hood, too, and can exercise himself, roll without getting particularly dirty, enjoy the space and liberty of the field and the company of other horses, and be far happier than his fully stabled colleague. Such horses are more relaxed, suffer less from general health problems – especially respiratory ones, as long as the turn-out time is significant – and fitter, as they can take that slow exercise which is so essential. Nor are they such a burden on their owners who, however, still have the advantage of being able to clip, corn-feed and groom their working horses.

The horse will be fitter than a fully grass-kept one, and certainly cleaner and easier to groom, as he will not need to carry his full winter coat unless the owner wishes him to do so. He will not suffer from exposure and does not, therefore, put much of his food into keeping warm but can put it all into his job.

This system can be used for Shetlands and racehorses alike and everything in between no matter what their work level or diet, and by using it you'll make your own life easier and your horse's healthier and happier.

YARDING

Here again, the essential principle is that the horses are not cooped up in their stables – though certainly, they do not have as much room as horses out on pasture. The yard area can vary greatly, but the usual system is to have part of it covered so the horses have shelter, and part open so they can be outside. Alternatively, the system can consist of large indoor yards or barns in which the horses can wander at will, with large double doors leading to the field, and these can be left open or closed as you wish; this is the ideal arrangement.

The flooring can be sand, pulverised bark, wood chips or anything similar, and the ground beneath simply natural earth, chalk, rammed shale or any other material which will provide natural drainage; then all the maintenance that is needed is regular removal of droppings and excessively dirty bedding.

This system is much more natural for the horses, who as a result remain more content and relaxed. As in any system where horses are mixed together, you must just be careful that you do not put together two or more animals who are aggressive towards each other. Large hayracks can be put round the walls, and a communal water container provided as in a paddock, and maybe long troughs for concentrate feeds depending on the system being used. Yarding saves a great deal of time, and the accommodation is cheaper to provide than, for example, a row of loose boxes to accommodate the same number of horses.

ACCOMMODATION AND FACILITIES

Basic facilities were mentioned in Part 1, but let's go into these in a little more detail now.

Stables

Most horses and ponies these days are stabled in loose boxes (called box stalls in the USA), where the horse is loose and can move around freely in his box. Standing stalls were always narrow, with horses tied by the head so they could lie down and get up but could not turn round properly. They are still used for some categories of horse which are given a great deal of exercise, for example some police and military horses and delivery horses, but they are not suitable for domestic, privately owned horses on comparatively restricted work.

Boxes can be square or rectangular; with the latter you must ensure that the horse can stand along the narrow width with at least a foot in front of and behind him, otherwise he will not be able to turn round safely and will feel restricted – and this is really not enough room. A box of about 12ft by 10ft is suitable for a horse of 15 hands high; 12 ft sq for a 16 hand horse; and for ponies, say 6ft by 8ft for a 13.2 hand pony, or about 8 ft sq for a 14.2 hand animal. The sizes are approximate, but you should not have anything smaller as the animal must be able not only to turn round, lie down, roll and get up, but also lie flat out to sleep – all in comfort and safety.

As for height, 7ft 6in to the eaves of a ridge-roofed stable and 10ft to the lowest part (usually the back) of a single-planed roof, should be regarded as minimal but *many* stables are lower than this, which can be dangerous if the horse is inclined to

Yarding is an excellent method of keeping horses, saving a great deal of mucking out time, giving the horses company and space to move around. The doors can be left open, as desired, so the animals can go out to pasture or stay indoors, as they wish. This method is suitable for hard-working horses as well as breeding stock or resting animals. Stallions also benefit from living, albeit singly normally, in similar pens with free access to pasture, as required.

Because of the open boarding the ventilation is absolutely ideal.

STABLE BYE-LAWS

Stables have to comply with certain bye-laws concerning health and safety, involving both drainage and fire precautions, so check with your local council if keeping your horse at home. Otherwise, your stable proprietor will have done this.

LIVERY

Home versus livery

Most owners would love to have their horses at home, but for most this is simply not practical as they do not have the facilities to do so. Taking into account certain laws and other local conditions, it is often quite possible to keep a horse at home with a stable in the back garden, perhaps turning it out in nearby rented grazing daily to enjoy the freedom and company of others – provided nearby grazing can be found, of course. Even this is not essential provided the animal can genuinely be given enough exercise, and perhaps be allowed to potter about his back garden (which in this case couldn't *be* a garden, of course).

At the time of writing, you are permitted to have a stable on your own property and class it as an enlargement of the property provided that it does not exceed one tenth of the cubic content of the original house, or 50 cubic metres subject to a maximum of 115 cubic metres. Check this with a solicitor in consultation over the deeds of your property (borrow them back or get a copy from your mortgagor, if necessary), and check on any local planning regulations, too, if you really want the horse outside your back door.

Remember a few other things, however: although this is by far the cheapest way of keeping your horse, it does have the disadvantage of being extremely tying, and when you are ill or away you'll probably have to put the horse out at livery unless there is someone living in who will look after him. It is extremely risky to leave him at home and just get someone to call in, even if it is several times a day; the horse could be stolen, or could fall ill or injure himself and be left for many hours before he is attended to. Also, if the attendant is not 100 per cent reliable, or is herself prevented from coming, again the horse could suffer.

It is also necessary for you to be fully competent to look after him, or at least willing and able to call on expert help when in doubt – and to be clear when such help is needed.

For these reasons, I do recommend that for your first year of ownership you keep your horse or pony at a BHS or ABRS approved riding centre, where you will absorb a great deal of knowledge and experience of horse care during all seasons under all circumstances. You'll also have access to expert supervision and lessons on the spot, and can gain an immense amount of knowledge just by being in such a place and by talking to other owners, staff and instructors.

Full livery

Short of employing your own groom, this is the most expensive and least tying method of keeping your horse. The stable should do everything for you, and you should never actually *have* to go and see your horse except when you want to ride, if this suits you. Full livery costs vary greatly around the country depending on the facilities available and on local business rates, so shop around by ringing a few yards recommended by your adviser.

For full livery the horse should be fed, watered, groomed, mucked out, turned out and brought in if you wish, have his tack cleaned and rugs laundered and be properly exercised according to his state of fitness. You have to pay extra for extra schooling, and also for farriery and veterinary costs but the stable staff will attend your horse during these processes. If the horse is ill he will be nursed, but you should not have to pay anything more for this.

Some stables, however, do not include ridden exercise in the cost of full livery, only turning out, so check on this.

There are few sadder sights than a well-fed, rugged-up, spotlessly clean and thoroughly cosseted horse standing with his back to the door, miserable, unhappy, depressed and completely uninterested in his surroundings. When looking for a stable for your horse, or for a school at which to study equitation and riding, look first at the attitude of the horses and ponies as well as their physical management. Happy animals are far more important than everything done by the book with no imagination and no attention paid to the horse's state of mind.

Part livery

This means that the stable staff do some of the work involved in caring for your horse, and you do the rest, by mutual arrangement. A useful division for working owners would be for the stable to feed, water and muck out the horse and give him a light grooming daily, but for you to do thorough grooming, exercising and tack cleaning. This arrangement is obviously cheaper than full livery.

Some yards also make allowances in their charges for owners who permit their horses to be used in the school for training students or teaching clients; however, although this sounds fine in theory it sometimes causes problems in practice. It is never good for horses to have too many riders, particularly of novice standard, as it confuses them and their standard of schooling tends to deteriorate. Also, you'll often find that the school needs your horse most just when you do – at the weekends. Nevertheless you may be able to come to some arrangement, particularly if you work odd hours; for example, you could stipulate that only riders of a given standard use the horse, such as staff for escorting hacks or for demonstrations – it is worth exploring these possibilities.

Grass livery

As the name suggests, this means the horse is kept entirely at grass, so you'll need to check that the field really is properly fenced and – most important but so often lacking – does have truly effective shelter, maybe in the form of a man-made field shelter, and that the water supply is adequate. The ground should not be of the type to become poached or even waterlogged in wet weather, or as hard as concrete during a dry spell.

Some grass livery owners 'do' (look after) their horses entirely themselves but occasionally field owners will do some of the work, such as checking, giving supplementary feeds and so on; so check what services are available.

DIY livery

In this, you 'do' your horse entirely yourself (hence 'do-it-yourself' livery) and simply rent the stable and grazing; and if anything, this is even more tying than keeping a horse at home, as you have the time, trouble and expense of travelling to wherever your horse lives. The big risk in this sort of livery is that some owners are really negligent; however some premises – the better ones – do have a manager or owner who will supervise to make sure horses are being properly looked after, and in emergencies such as an owner's illness, will give the horse basic care.

rear in his box or if he is likely to do so should something in the yard outside startle him. A 16 hand horse rearing can hit the roof of a lower stable with his head.

There is another important reason for lofty roofs: air space and ventilation. Horses are very prone to respiratory diseases due to muggy air, airborne bacteria, dust and mites, and gas such as ammonia which is formed when urine and droppings decompose in an improperly cleaned stable. The sensitive linings or mucous membranes of the throat, windpipe and lungs become irritated and then provide an ideal inlet for bacteria and viruses such as influenza. Some horses are allergic to dust and to the mould spores which are present on even good samples of hay, and on straw where this is used for bedding, and bad ventilation compounds the problem. The horse often ends up with an allergic condition commonly called broken wind, the correct name for which is either small airway disease (because of the swollen and constricted state of the airways) or chronic obstructive pulmonary disease. The condition can be managed, however, by putting the horse on what is known as a clean air regime; this involves providing dust-free bedding such as dust-extracted shavings or shredded paper for bedding, and feeding thoroughly soaked hay or moist, conserved forages called hayage (sold under different brand names).

The ventilation of the horse's stable can help greatly to relieve and prevent such problems. Windows should be above head height, and although it does not really matter whether they open upwards and inwards or downwards and outwards, you do need them placed so you can create a cross-draught – but above the horse's head. If the top door is always left open (except when wind and rain are blowing directly in) your window can be on the opposite wall, at the back of the box. Windows are, however, usually placed on the front wall next to the door specially to prevent this, because most boxes are too low to permit windows above head height and the cross-draught would then be detrimental in inclement weather, although an advantage in summer.

Strongly built timber stabling which allows the horses to put out their heads and feel less closed in and also get fresh air should the boxes have limited ventilation. The windows here are slightly unusual, opening right down and outwards for extra ventilation. Unfortunately, as is often the case with loose box windows, the added light and air is partially blocked by the top leaves of the doors opening across them. Windows in the backs of the boxes, roof windows and also ridge-roof ventilators and/or louvres at eaves height all effectively overcome this disadvantage.

In addition to windows, louvres in walls or gable ends are useful and ridge roof ventilators are a real boon in any stable, preferably running the full length of the box, although this is still an unusual feature. This device is pictured and explained in the accompanying line drawing. In a stable which is large and high enough, it is quite possible to provide proper ventilation without creating an uncomfortable draught for the horse. Unless a horse is ill, it is far better to keep him in a cold-ish box which is free from low draughts, and wearing clothing for warmth, than in a dangerously muggy atmosphere with the door and ventilation outlets closed. Even sick horses need fresh air, however, so pay close attention to the direction of the wind and close any outlets on the side of the box from which the wind is coming, to prevent undue draughts.

One of the major faults in all stables is bad ventilation. Clean air is crucial to the health, comfort and performance of horses, whether working or not. High stables with air outlets above the height of the horse's head are an effective way to achieve this, and one of the best methods is to have either ridge-roof ventilators (protective roof-shaped cowls open round the bottom and often with louvres up the sides set above an opening in the roof), one per box, or a fully open ridge to the roof with a protective cowl, as shown here, all the way along the roof, the cowl obviously being supported at intervals along the roof ridge. With a single-pitch roof a full-length opening should be left underneath the overhang on the highest side.

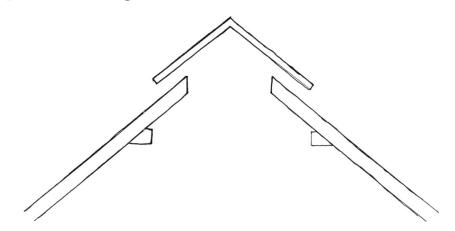

Stable floorings are usually made of concrete but these are really not the best. They are cheap but are cold and absorbent (and remember the moisture in stables is urine, not water). If used, they should be sloped slightly to some outlet near the drain outside the box (inside drains are dangerous and tend to block easily) and the floor should be roughened, and grooved to assist drainage. However in practice very little urine will be seen to run away outside: most of it is soaked up by the bedding.

A better floor consists of a top layer of laid bricks with coarse grit in between, all laid over hard core on a bed of earth, the earth for the stable site having been excavated to a depth of about a foot. Another is of loose weave asphalt, which is ordinary coarse asphalt laid in the normal way over pebbles on rubble but only lightly smoothed over, *not* tamped down or rolled, as flattening it will destroy the loose weave effect, fill in or squash out the holes and thereby nullify the drainage effect – the urine has nowhere to drain to and the whole effect is lost. Obviously the floor must be allowed to dry and harden thoroughly before the horses and bedding are put in, or the horses, too, will flatten it. Rammed earth or chalk also make simple but successful flooring materials.

Stables need to be strong as well as roomy, because a horse can easily pull or kick down a flimsy building. Double brick walls, or single brick lined with proper kicking boards (of tongued and grooved boarding or 5-ply wood panels) are excellent – warm in winter and cool in summer – but, of course, expensive. Solid concrete blocks or reinforced concrete is suitable, but better lined fully or partly with wood (the lower half to form kicking boards) to give better insulation; hollow ash breeze blocks are useless, however, and far too weak to withstand a good kick. Stone is cold all year round, and metal should never be used (eg corrugated metal sheeting) as it is unbearably hot in summer and very cold and damp in winter. Asbestos is similar, and too weak.

The roof should be of wood with roofing felt or felt tiles, preferably with the added insulation of an interior wooden lining. Some firms make specially insulated roofing materials which are good for any animal housing.

KEEPING THE STABLE DOOR CLOSED

Stable door bolts must be really heavy-duty, and those normally used are specially designed ones of galvanised metal. The top bolt can be secured down by clipping a trigger-type lead-rope clip through the hole which will stop the horse learning how to undo it; and the bottom one, the so-called 'kick-over' type, is specially designed to be turned over with the foot when you have both hands full. However, it only works if the door fits perfectly and is not warped! The reason for the two bolts – apart from added strength and security – is that should the horse lie down and push against the bottom of the door he will not press it open, maybe getting a foot in the gap and seriously injuring himself.

Wood is also very commonly used for stabling, and provided it is at least half lined it can be good if roomy. Badly ventilated, small wooden boxes can, however, be very muggy in summer and are not recommended then.

Windows can be of glass with an inside guard to prevent the horse breaking them, or of reinforced glass or PVC.

Stable fittings include hayracks for holding hay, the best type being of strong galvanised bars or mesh designed to fit across a corner of the box – on a flat wall they would project and the horse could knock his head. They should be fixed (one per box) at horse's head height. Hay can also be conveniently fed in haynets (like giant tennis ball bags), hoisted up (see Part 4) and tied to a strong ring or bracket, again at head height so they do not sag down and risk the horse getting a hoof through the mesh.

Automatic watering machines are available for piped supplies and should also be fixed in a corner at about the height of the horse's breast. The best are the type with a recessed plug in the bottom, to facilitate regular cleaning out.

The best mangers are stainless steel or galvanised metal, although polythene ones are also good, and cheaper. All mangers should be the corner type, fixed at the same height as the waterer; some have bars across the corners to prevent the horse scooping out his feed and wasting it. It should be possible to lift them out easily for cleaning, though if they are the fixed type they can be tackled with a damp scrubbing brush and cloth. Older stables often have lovely strong porcelain or earthenware mangers which have to be cleaned in this way.

Water is easily supplied in buckets or small plastic dustbins in a corner if your stable does not run to automatic waterers. They should be of semi-rigid or soft plastic or polythene (metal ones can be dangerous if the horse lies on them in the

WOODEN AND COBBLED FLOORS

Wooden floors are most unsuitable; they are soft and slippery when wet, splinter and wear badly, smell and harbour germs. Old-fashioned flat cobbles still occasionally found in old stables are, contrary to popular opinion, quite alright if kept well covered by bedding to disguise their slightly uneven surface – but check the drainage of the stable by pouring a bucket of water in the middle before installing your horse. If it drains away, fine, if not it's no use as your bedding will be constantly soggy.

ELECTRICITY, PLUMBING AND DRAINS

It is quite possible to care adequately for your horse without an electrical supply, but it is more difficult and not so pleasant; what you have depends on the age, design and former use of your stable and yard. Hot water is hard to provide without an immersion heater unless you have an Aga or suchlike in the tack room, but for a small amount such as to mix in a feed to warm it up on a cold winter's night, you can take a large vacuum flask to the stables filled at home with very hot water. Washing rugs and so on will have to be done at home (or at a launderette, if you can get one to accept dirty horse rugs!).

Lighting can be adequately provided with a supply of double-sided motorists' lanterns with a top handle, operated by battery. These can be hung on recessed hooks on two sides of the stable, and cope quite well. The same type of lamp can be hung in the feed room, tack room or wherever they are needed and as long as you keep a supply of batteries in the tack room drawer you'll manage well enough. And many's the time, on dark winter mornings and evenings, I have mucked out by the light of my car headlights!

Proper wired-up electricity is obviously much more convenient, but it does present a risk. Stables and their ancillary equipment, fodder and so on, are areas of extreme inflammability and your supply should be checked yearly. Old wiring plus the inevitably slightly humid atmosphere of a stable can spell death, as the wire coverings gradually rot and bare wires will become exposed; light bulb fastenings are also affected, and should fittings be within reach of a horse you can bet he will find them, damage them and perhaps cause a serious fire – and maybe give himself a nasty shock, too.

All electrical wiring should be run inside galvanised metal conduits to prevent interference from the horse or rats. Light fittings should be high up in the ceiling of the stable and should preferably be concealed behind toughened glass. Switches should be the waterproof type to withstand rain even if under a verandah, and should be on the outside wall of the box out of reach of craning necks. Yard lights should be the waterproof, reinforced bulkhead type.

If boxes have automatic waterers, piping should, like wiring, be run inside metal conduits to prevent horses prising them away with their teeth for fun, causing a deluge inside the box and cutting off the other horses' supply. Although in Britain we have had very mild winters of late, it is still best to lag thoroughly all the piping to stable and field supplies which is less than a foot below ground; this should prevent freezing up, cut-off supplies and subsequent bursts.

A tip about hosepipes: don't leave your hosepipes outside in freezing weather – if they freeze inside you won't be able to use them and the expanding ice could split them, making them useless even when thawed out.

Drains in stable premises are very prone to blocking up due to bedding and fodder blowing into them; they should be covered with heavy duty mesh covers and cleaned superficially every day. The covers should be firmly fixed so that horses walking about the yard cannot get near them and put a leg down them; not only would this severely frighten the horse, but it could cause an accident as serious as a broken leg as the horse struggles to right itself – back injuries are also caused this way.

Old-fashioned centre drains in stables are especially prone to these two faults; they should be blocked off completely and diverted if at all possible.

Some American barn systems of stabling, with rows of loose boxes sited inside large buildings, have drainage channels with perforated metal covers running along just outside them. The Irish National Stud has a flush system for one such barn – flush loos for horses! Most barns do not, however, but it is just as important to flush out these interior drains as it is the outside ones: use very hot water and washing soda, not caustic soda – unless the horses are out and the resultant obnoxious fumes are given ample time to disperse before their return. Caustic soda is not suitable for plastic drainage pipes as it burns them away quite quickly.

night) and can be slotted into fold-down holders (folded down when the bucket is out being cleaned or filled, to prevent the horse getting a leg down them) fastened to the wall in a corner near the floor. They can, of course, be placed on the floor, free-standing, for quiet horses who don't knock their containers over — but these seem to be a minority commodity!

Storage facilities

Hay and straw are bulky and need barn storage unless you are going to buy frequent small loads, which is not the most economical way of buying these things; though many owners without large barns have no choice. They should be kept dry, well-ventilated, and downwind of the stables and house, not only to prevent any spores blowing around, but also in case of fire. Open-sided barns should not be used as the bales will inevitably get soaked when it rains, which means the outside bales will be largely ruined.

Concentrates should be stored preferably in metal, vermin-proof bins in a cool, dry feed room (a spare stable or simple shed will do), although many owners make do very well with metal dustbins or tough plastic ones (except the latter are not rat-proof). If you do have to store your feed in the sacks in which it is delivered, make sure they are off the floor a little way in case of damp, and that the shelf is high enough to permit a cat or terrier to get under.

Ideally the feed room should have a water supply and a kettle (if no hot water) so water can be boiled to clean feed containerrs and damp ordinary feeds, melt black treacle and so on. The old practice of cooking feeds has been largely superseded by modern horse feeds, although some stables still boil barley and linseed and make bran mashes — although bran mashes are not now recommended by equine nutritionalists (see p113).

Tack should be kept in an aired, dry room if it is to remain in good condition. The plastic-covered saddle racks shown here are good as they do not reach into the saddle gullets and stretch the seats from underneath. The bridle racks, being semi-circular, are better than pegs, and certainly better than a nail in the wall, as the bridle headpieces will not develop a bend or crack in the top. The racks and shelves on the left keep boots, grooming kit etc. tidy.

Tack and clothing are usually stored in a tack room which is separate from the feed room, as these tend to be dusty places which does not help either the cleanliness or appearance of tack. Nor are steamy feed rooms good for leather, which is best kept in a dry atmosphere of about 20°C or normal room temperature.

Saddles are very expensive and should receive the greatest care. They should be kept on racks which do *not* press up into the gullet as this stretches the seat from underneath, and are best covered by a saddle cover or old towel. Bridles should be hung on semi-circular brackets (old saddle soap tins nailed to the wall are excellent, and *free!*), as hanging them on a narrow peg or nail in the wall stresses and cracks the leather irreversibly; bridles, too, are expensive. The reins should be hung up with the bridle headpiece or caught up by the throatlatch to prevent their being trodden on or tripping someone up.

The tack room should also contain cupboard and/or drawer storage for numnahs, boots, bandages, tack-cleaning gear and often grooming kit, and shelves, chests, baskets or special hanging racks for clothing. It is especially important to have something fixed up in a warm place for drying turn-out rugs, as these will never dry fast enough in normal winter temperatures. Old-fashioned clothes' racks firmly fixed to the ceiling, with a pulley and rope to haul them up and let them down, are a real boon for drying anything wet or damp, as tack, bandages, boots and clothing often are. They carry a lot of equipment high up near the ceiling, taking advantage of risen warm air, and can be used in the tack room or kitchen for coping with wet riding clothes as well as horse gear.

Ideally, tack rooms should be fitted with a thermostat to keep them at a constant temperature. The room just needs to feel aired, not actually warm, or even hot which will dry out and crack leather, so the cost of this facility is not excessive especially when compared to the high price of spoiled tack.

A water supply is important here, too, for tack cleaning. A saddle horse (as illustrated on p139) is helpful, and so is a bridle hook which holds three bridles for cleaning at head height – but just because of this it is most important that you remember to take it down after use, otherwise it might cause serious injury to head or eyes by people accidently walking into it.

Field and fencing

The grass in your horse's field need not be lush and long, as horses do best on poorish quality grazing as long as it is clean. Contrary to popular opinion however, they will eat long grass if they find it palatable.

If you want grass to form a significant part of your horse's diet so that you can save on other feeds, you will need about two acres/1 hectare for the first horse, and at least one acre/.5h for each additional one, otherwise the field will not be able to provide a great deal of grass and will become primarily an exercise area. There is nothing wrong with that, of course, provided the ground is not the type which gets bogged into an oozing morass in winter, or bakes hard and cracks in summer. Ideally, you want lightish, loamy soil, well drained, with a gentle slope to some outlet such as a pond or ditch; and both of these should really be fenced off for safety as horses – who *will* find trouble if there is any – may fall in and be unable to get out. This is especially important with ponds in winter, when horses may try to walk on the ice.

It's not easy for those with little knowledge to assess pasture, but look for extremes of type, such as ground which has obviously cracked at some point (probably clay land which is far from ideal for horses); for spiky tufts of rush-type grasses (which indicate marshy conditions); and excessive weeds such as docks, nettles, thistles or poisonous plants, particularly ragwort, all of which indicate unsuitable and/or neglected land.

Ideally there should be a sturdy field shelter with an open front sited on the

NATURAL SHELTER

If there is no shelter shed, natural shelter is essential for horses out for more than just a few hours a day. There should be high, thick hedges on the northern boundary at least and also against the direction of the local prevailing winds, and preferably on all sides of the field, and there should really be significant clumps of trees for shade and overhead shelter. However in winter, broadleaved trees will offer little protection from rain as they will have lost their leaves, hence my recommendation for a proper shed, the size of which should be the size of a loose box for one horse and half as much again for each additional horse. In fields with large herds, domestically speaking, of more than about six horses, two or more smaller sheds are better than one large one (unless the horses have access to a covered yard from the field) as arguments can occur and some horses will be kept out of the shelter. Incidentally, rain dripping from tree leaves on to sheltering horses encourages skin infections such as rain rash.

Natural shelter in the form of good, thick hedges, particularly on the windward side of a field, and overhead shelter from dense spinneys or woods, is excellent but comparatively rare. Field shelters such as this open-fronted shed are also excellent provided the entrance is high, wide and welcoming and it is not dark inside nor the approach boggy, stony or rough. It is also important to ensure that no horse is threatened and frightened of entering due to a dominant field companion, as here. If this is happening alternative companions must be sought and either the dominant horse removed or those prevented from using the shelter (normally the former course of action works best).

Hawthorn

Midland Hawthorn

highest part of the field to prevent the inside becoming boggy, and with its back to the prevailing wind – if there doesn't seem to be one, then face the entrance south-east to protect the inside from Britain's generally south-westerly winds.

Fencing is an important topic: not only is it necessay to keep horses within its boundaries, but it must be done safely – and some fencing is anything but that. Barbed wire is commonly seen for horses, particularly where they are grazed with cattle, and it's true some horses do seem to graze safely for years inside it; but it really is the most dangerous of all. I have seen it scar, injure and lame horses for life, and have personally known of several horses put down because their injuries were too bad to put right. Horses in a field often do what comes naturally – they play and gallop around and, sadly, do not always seem able to see where they are going clearly – or to concentrate, at any rate, often ending up on the fence. If that fence is barbed wire the results can be catastrophic, as you can imagine.

The ideal fence is thick, high, prickly hedging (hawthorn is a traditional favourite); the best man-made fencing is the type of post and rails illustrated on p43. Other good fencing materials – most of which are cheaper than wooden posts and rails – include plain, heavy gauge wire strained really taut on wooden posts (the commonly seen concrete posts are not really safe for horses); flexible PVC rails on wooden posts; diamond mesh fencing, usually galvanised or plastic covered metal, inside a sturdy wooden framework; or very small mesh sheep fencing (although it is difficult to get it small enough to prevent horses' hooves going through it) on wooden posts or frames. Nylon strand fencing is also now coming into use.

Any other fencing will probably be too weak or too dangerous for horses. Ideally you want to present as smooth a barrier as possible so they do not get cut or scratched, but one with a certain amount of resilience so that severe contact injuries (as can happen with, for example, concrete posts) are minimised.

Fencing should be as high as the withers of the tallest horse to use the field, and preferably higher.

The most convenient way of providing a water supply to a field is by piped water to automatically filling troughs, although for horses, the troughs must have no sharp corners or edges. A pressure fill or a conventional ballcock device can be used, or simply a tap operated by hand on daily checks. All mechanisms should be covered with a wooden box to stop the horses interfering with them. Don't make the mistake either, of throwing down a load of rubble around a static trough in an attempt to counteract the liquid mud which can form round it in wet weather, as is seen on many farms. Cattle themselves are reluctant to pick their way over such a surface as it is extremely uncomfortable to their feet, and horses often simply refuse to do so, sometimes becoming dangerously short of water as a result. The siting of the trough can avoid this: it should be on a high, well-drained part of the field and preferably with a clear space around it for safe movement. In practice, one trough often has to serve two fields and so is often on the boundary fence, either set along it or crossways to it. The latter positioning can be dangerous because it projects, and horses galloping along the fence line may bump into it.

Fields without piped water can have containers filled by hosepipe – large plastic dustbins rammed into tractor tyres and tied to a fence post are a favourite. If there is no tap within reach of the field, water must be taken by means of a water bowser. Remember that ponds are often stagnant and polluted and have unsafe approaches, and the same goes for streams: in this case an artificial supply is needed.

Gates and sliprails

These take a lot of hard wear and tear, and must be really strong and well hung, or safely fixed. Gates which stick on the ground or which don't swing shut naturally are a real menace; you should be able to open a gate with one hand while carrying feeds or leading a horse, and when you let go it should swing shut of its own accord so there isn't an inviting gap for other field occupants to nip through and off.

Strong wooden barred gates are excellent, as are the ones made of strong tubular metal; however both of these should have their bottom halves filled in with strong metal mesh so that if there are horses larking about near the gate they cannot get their legs through between the rails. They should have recessed bolts which the horses cannot fiddle with easily, and should really be padlocked at both

It's always best to turn horses out with no headcollars on, but this is not always feasible, especially with horses who are hard to catch or difficult to identify, such as plain brown Thoroughbred yearlings! For this reason breeding stock at least, epecially when away at stud or boarding out, traditionally wear headcollars with their nameplates on.

It is almost impossible to prevent gateways and areas receiving hard use from becoming poached in wet weather unless they are excavated out and replaced with hardcore under sand. Used bedding from the stables can lessen the problem and can also be spread on the areas when they become hard, dry and rough as the ground dries out, as here.

Wheat straw – still a popular bedding material.

Wood shavings, probably as common a bedding material now as straw, if not more so.

ends when horses are in: this should prevent easy access by horse thieves, and thwart any horse who does actually manage to manoeuvre the bolt, the first point being particularly important where fields are accessible from public roads or rights of way.

Sliprails are a cheaper alternative to gates and work well, provided one end of each rail is bolted through its receiving post and the other is slotted into a closed-top holder to stop horses getting their necks under the top rail and jiggling it along till it drops so they can get out and explore. This is a natural tendency for horses which is, unfortunately for them, frowned upon in our society! This doesn't stop them trying, however.

With all gates, sliprails and fencing just remember that everything should be smooth with no protruding nails, and no sharp-ended bolts or jagged screws, fastenings or metal – then you won't go far wrong. Don't be tempted to install fancy hunting-latch type gates with handles sticking up above the top rail, meant to be operated from horseback: they positively invite headcollars to become hooked round them, resulting in at least one badly frightened and maybe injured horse.

BEDDING

The main requirement horses have of any lying-down place is that it should be dry; they are also seen to avoid rough areas, although when rolling (not the same as lying down proper) they will often pick the muddiest patch they can find. We provide horses with bedding for various reasons: to encourage them to lie down and rest, to help keep them clean and warm, and to protect them by providing a resilient cushion for legs and body. Bedding must not present a health hazard to horses or irritate them in any way, and from our point of view it must be reasonably economical provided the foregoing needs can be met.

Materials
Here in the UK the most common bedding materials are straw and shavings, their use being vastly greater than sawdust, peat or shredded paper. Straw is traditional, warm and easy to work with but, as already mentioned, can cause allergic reactions in more than a few horses. Shavings are cheaper but not as warm, and they are fairly easy to work with once you get the knack. They are not as stable as straw, and scatter easily till the bed's foundations are established (which means keeping a bottom layer of damp shavings). Sawdust can be just that – dusty – unless damped down slightly; and peat, despite teachings to the contrary, really *is* dusty. It gets everywhere – up your nose, in your hair and clothes and down your boots – and if it does that to *you*, who come into relatively rare contact with it, just think what it is doing to your horse. If it gets up his nose it will be doing so at least twenty-two hours a day for a stabled horse, and up his nose means down into his lungs where you certainly do not want any dust. Peat is very absorbent, particularly moss peat, usually damp and, therefore, cold in winter and humid in summer. It even freezes. Needless to say, I do not think it makes good bedding material.

Shredded paper on the other hand is certainly one of the best bedding materials available. It is marketed by several companies, the best of whom claim it is practically sterile with no dust or spores; because of this it is recommended for any horse susceptible to wind problems. Athletic performance horses, too, are advised to use shredded paper bedding because they need their lungs to be working at full capacity; a challenge such as that presented by dust and moulds uses up the horse's energy resources in fighting it off, even if symptoms of allergy or disease never actually show in the horse. Shredded paper forms part of the clean air régime recommended for such horses; or indeed for any horse whose owner is concerned for his physical condition.

Shavings and straw can now be purchased 'clean', however; they are usually described as 'dust-extracted', having been put through a commercial vacuum machine which sucks out over 90 per cent of such material into a disposable bag. Machines *are* available for purchase, but at the time of writing cost over £1,000 so are possibly only practical for large yards, and are currently only suitable for hay and straw, not shavings. Obviously, this process is not possible for sawdust or peat which are little more than dust themselves.

This wheat straw bed is thick in the middle and has been banked well up round the sides for extra comfort and to help 'bounce' the horse back into the centre when he lies down or rolls.

Managing a bed

The object is to keep the bedding material dry, clean and thick. If you have proper drainage flooring (as already described) the urine will drain through and away, keeping your bedding much drier than concrete floorings which waste vast amounts of material. Otherwise you will have to remove not only the droppings but the wettest, dirtiest bedding in fairly large amounts every day. This is the *full mucking out* system: first you pick up the droppings and put them in a wheelbarrow which you can place across the doorway of the stable to stop the horse escaping if he is in and untied. You can tie him up, of course, but it's much easier to muck out, if possible, while the horse is out.

When the droppings have been removed, choose the cleanest corner and shovel or rake all the clean bedding into it, removing any further hidden droppings as you go. Next shovel out dirty, wet bedding, scraping the floor of the stable with the back of a shovel for any stubborn residues, and then ideally, sweep and rinse down the floor. If possible, leave the bed up for the floor to air and dry, preferably while the horse is out. If you use a different clean corner every day for piling up the clean material, the whole box will be regularly swept. Pile any half-clean material in a different corner.

To *bed down* again, scatter the half-clean bedding evenly over the floor as a foundation, then layer the salvaged clean stuff over it in another layer. Finally

Wood chips (with pencil as size guide) used for surfacing exercise areas. They are too hard, heavy and uncomfortable to use for bedding. Compare them with shavings bedding shown left.

55

CARE AND MANAGEMENT
Mucking out

MUCKING OUT

Mucking out is time-consuming, heavy work. Make the job easier by using a large barrow which takes a lot of muck and cuts down your trips to the muck heap and which, having two wheels like this one, is easy to balance, steer and push, cutting down the energy needed to manipulate it and also the chances of your spilling your load on the yard, creating another job sweeping it up again. Small, single-wheeled barrows should be abolished!

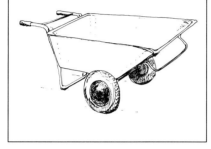

A selection of mucking out equipment. Leaning against the wheel-barrow are, from left, a yard broom, five-tined fork, nice big shovel, a plastic droppings scoop and a shavings rake for use with it.

This shavings scoop is ideal for removing droppings from shavings.

56

bring in fresh bedding – this provides a third layer on top, but is mainly to bank up round the sides of the box for extra comfort, warmth and protection, and also to help prevent a rolling or recumbent horse getting stuck against the wall with his legs arranged in such a way that he cannot get up again (this is called becoming 'cast', and can cause horses to struggle and panic with sometimes serious external and internal injuries). The banking should help him bounce back into the centre of the box so he doesn't get cast in the first place.

Another method of doing a bed is called *deep litter*. In this system, only the droppings are removed several times a day; the bedding itself is largely left to consolidate into a thick cushion which gradually rots down from the bottom, a little fresh bedding being added to the top daily as required. It sounds horrible, but in a *well drained, well ventilated box* it can work well. The secret is to be meticulous about removing droppings (difficult overnight) as otherwise the bed turns into an indoor muck heap; the accumulated bed of deep litter is removed when, say, the bed in the centre of box (where it gets most wear) is significantly over a foot thick. I have to say it does work better in winter than summer!

Semi-deep litter is a compromise between the two systems. The droppings and very worst bedding are removed two or more times a day, the clean stuff from the banking used to fill in the gaps, and new material banked round the sides. This method works well for working owners who don't like deep litter but who only have time to muck out fully at weekends. You could clear out fully every month or so, as wished.

Setting fair is used to describe the minor attention given to a fully mucked out bed during the day and evening. Full mucking out is normally done in the mornings to get rid of the night's efforts (horses pass about eight piles of droppings in 24 hours, sometimes more). Especially in the evening, any further droppings are removed, the bedding fluffed up and perhaps a little fresh material added to the middle for the night.

MANURE

The general feeling is still that straw makes the best manure for gardens or nurseries, as well as for farm use. It adds valuable nutrients and humus to land when combined with droppings, unlike shavings which deplete the land of nitrogen unless very well rotted (more than a year old). Peat bedding, where used, makes good manure but turns soil acid over a period of time, which is not conducive to healthy growth in many garden and nursery plants – apart from rhododendrons and azaleas which love it, along with a few others. This is not a reason for using peat as bedding, however.

Much is said in old books about keeping the muck heap excessively tidy, and that beating it down and squaring it off makes for better rotting down. In practice the difference this makes is small and, anyway, nurseries like fresh manure usually to mix with their own compost activators. Shavings manure needs special nitrogen additives added to it, but so many yards now use shavings that the nurseries which they supply have no choice, in some areas, but to take your shavings and add the nitrogen themselves.

Just keep your muck heap reasonably tidy, perhaps by partitioning it off with old doors, used timber, asbestos or metal sheeting or whatever you have available, and let your contractor take it away by arrangement for a modest payment to you. Many people simply back their muck heap up to a building, but remember that it will rot wood, and because it smells it has to be kept downwind of and not too near stables or houses; however, it must be within a reasonable distance as it will be in frequent use.

Many contractors will leave a container for you to put your muck directly into: they remove it when full and leave you an empty one. Site it in a dip if possible so all you need do is tip your muck effortlessly down into it, or construct a gently sloping ramp and platform – pull your barrow behind you up the ramp (much easier than pushing it up in front of you), turn it and tip the muck out. Heaving muck up out of a barrow with a shovel into a container which may well have sides four or five feet high is no joke.

Mucking out into two piles may sound like a lot of extra work, but in fact it isn't and it does have definite advantages. Make one pile mainly droppings, and the other mainly dirty bedding; you'll find the droppings pile will make a more easily saleable and valuable manure (being more concentrated but still containing humus, and the dirty bedding pile is ideal for use around the premises – it can be put in a thick layer in poached areas such as gateways, round troughs, along fence lines, under trees and in shelter entrances, all of which will never grow much grass anyway and form only a small part of your pasture area.

In urban areas you can simply put your muck direct into plastic sacks and sell it at the gate, or even give it away to gardeners, so you don't have neighbours complaining about your Horrible Heap.

As long as it is very well rotted (over a year old and preferably older), horse manure can be used to spread on your own paddocks as part of your land management routine – fresher manure taints the land with its own smell and repels horses, who normally hate grazing over their own droppings. In the older muck heap any residual parasite eggs and larvae will have been killed off by the heat generated in the heap itself, so will not reinfest the land. If you are going to keep your muck for this purpose, cover the pile when complete as too much rainwater will hinder the rotting process. However, air is essential, so provide ventilation from the sides, and get the heap turned inside out after about six months – a Herculean task by hand, so consider hiring or cajoling someone who has a mechanical shovel to do it for you.

Pick out the feet before and after every ride. Work from the heels forward, like this, to lessen the chance of pushing grit under the loosest part of the shoe at the heels.

Hold the tail by the end of the dock and brush out a lock at a time, gently to avoid breaking hairs.

Skipping out is the term used for removing droppings only, and is so called because traditionally they are shovelled into a shallow container called a skip or skep. In the old days they were of basketware but now they are usually rubber or plastic, and upturned dustbin lids make good skips provided you don't later use them to cover your feed bins! Plastic mesh laundry baskets are also good.

Bed management is largely a matter of common sense. It doesn't really matter exactly how and when you remove the muck as long as you do so properly and at least twice a day and preferably more. *If your stable smells you are not doing the job properly and/or your stable is not properly drained and ventilated.* Rotting organic matter such as we are talking about gives off ammonia, and standing in it causes the horn of the horse's hooves to soften and disease to develop – this is commonly called thrush, a fungal disease of the feet; the hoof horn may actually disintegrate over a period. Ammonia fumes also cause respiratory problems.

Keep your horse's bed clean and dry, and thick enough for warmth and comfort and you will go a long way to keeping him healthy. But how thick is thick enough? A reliable practical test is to judge it by the length of the tines (prongs) on your mucking-out fork (illustrated in the accompanying photograph): for a straw bed, push the tines through the straw and if you can touch the floor the bed is too thin. For other materials which scatter more easily than straw, the bedding should cover the tines, although without the cushioning effect of straw you'll still reach the floor except possibly in a deep litter bed. When you walk on the bed yourself you should feel almost as though you are walking on a firm mattress without being conscious of the floor beneath. Finally, if you sit on the bed for half an hour can you feel damp striking through your jeans? If so, so can your horse – the bed is too damp.

GROOMING, CLIPPING AND TRIMMING

It's a lovely feeling riding a glossy, clean horse whose coat ripples with health and cleanliness. Wild horses, and those in a paddock having a long rest, particularly in summer, have a different sort of health and cleanliness but in their more natural condition, rain and mutual grooming keep their skins and coats clean enough for their purposes. Hard-working, stabled horses need extra help: they are not rained on very much, and because the skin is important for excreting many of the toxins produced during work and in the burning up of food, we need it in top working order to cope with our extra demands on it.

An in-work horse is easier to keep clean in practice than an idle, unfit one. Stabled horses need a thorough body brushing just about every day, and a full grooming will go something like this:

First, the feet are picked out, as shown in the photograph. Then dried mud and manure (often delicately called 'stable stains') are carefully brushed off, perhaps with help from a plastic or rubber curry comb. Then the body brush is used to get through the short summer or clipped winter coat, although it also has a reasonable effect on the unclipped parts of a stabled horse's winter coat.

Lean on your stiff, slightly bent arm and, using your weight, push the bristles through the coat as you brush in long, smooth strokes down and back, ending with an outward flick. Six strokes on one place should be enough. Every two or three strokes, draw the bristles, downwards, firmly across the metal ridges of the curry comb to clean them, and occasionally tap the comb on its side near the door – or outside, if the horse is tied up – to dislodge the dirt.

Formulate a routine so you do the same parts in turn so nowhere is missed. You could start at the head (but only as firmly as your horse will allow), followed by the neck, shoulders and breast and down the front legs and under the breast, then the withers, back, sides and belly, and finally the quarters and hind legs. Do both sides (obviously!) and finish off with the mane and tail: not many manes lie obediently on the correct, *ie* off-side of the neck, so push the mane over and,

GROOMING KIT

Here is a list of the main items you'll need to groom a stabled horse effectively:

*Hoof pick, for cleaning out the under-sides of the feet. The most important item in your kit.

*Dandy brush, with fairly long, stiff bristles used for removing dried mud and manure stains from the body and, if you're very careful not to pull out hairs, on the mane and tail.

*Body brush, with short fine bristles for removing grease, dirt and natural dandruff from the coat and for grooming mane and tail.

*Metal curry comb, for cleaning the body brush, *not* the horse.

*Two sponges of different colours, one for eyes, nostrils and lips and one for sheath, udder and general under-tail area. Never mix them up.

Other items you will find useful are:

*Water brush, like a smaller, softer dandy brush, for damping down (called 'laying') the hair of the mane and tail to encourage it to lie flat.

*Plastic and/or rubber curry combs for getting caked-on mud off the horse. Use carefully on bony areas such as head or legs.

*Stable rubber, just like a tea towel and used slightly damp for giving the coat a final dust or polish. An old silk scarf puts more of a shine on the coat.

*Hoof oil and brush, rather like a paintbrush. This gives a finishing touch, although some types are not absorbed and actually pick up dirt and bits. Unless the oil contains lanolin or glycerine it is unlikely to help the condition of the horn, either.

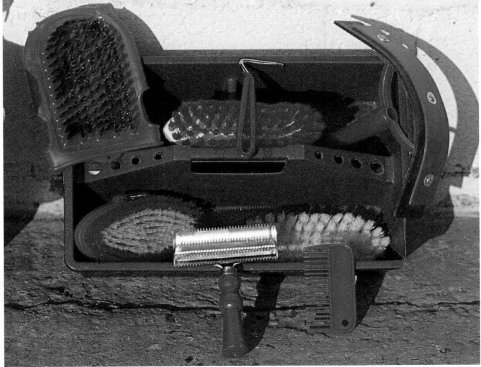

Plastic grooming tidies hold all your kit in one place. They often have trays or sections for plaiting and trimming equipment.

A selection of grooming kit. From top left clockwise, dandy grooming mit, dandy brush with hoofpick on top, sweat scraper, water brush with mane comb, small-type metal curry comb and body brush.

PICKING OUT FEET

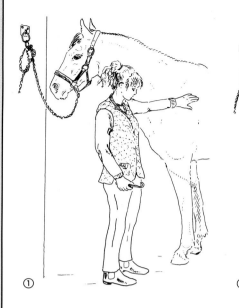

①

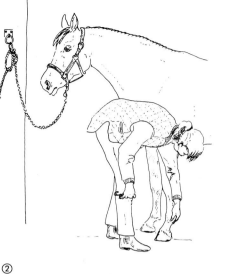

②

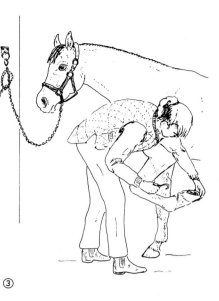

③

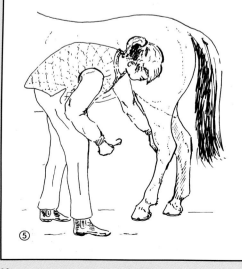

④

① To pick up a foreleg first stroke the horse on the shoulder . . .

② . . . then run your hand down the back of his leg to the fetlock. Grasp it and say 'up', perhaps pushing the back of his knee with your elbow and leaning your weight on his shoulder should he prove a little reluctant.

③ Hold the foot at the toe with your fingers. This prevents the horse leaning his weight on the leg, and you, as his fetlock would then be uncomfortably flexed and unable to bear his weight should he try leaning.

④ To pick up a hind foot, stroke the quarters first . . .

⑤ . . . and run your hand down the inside of the leg . . .

⑥ . . . to the fetlock, like this. Grasp the fetlock and pull it upwards saying 'up' at the same time and possibly leaning your shoulder against the horse to shift his weight over and make it easier.

⑦ Again, hold the foot with just your fingers on the toe or wall to stop the horse leaning on you. When putting down a foot, fore or hind, do just that – *put* it down rather than letting go of it as some horses will then wave it around and may kick and stamp, inadvertently hurting you in the process.

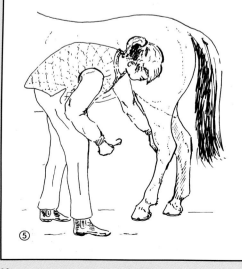

⑤

⑥

⑦

working from the roots unless the hair is long and tangled (in which case undo the snags with your fingers), start at the top and bring one lock at a time over to the right side with the body brush. Then do the forelock, and then the tail: grasp the dock right round near the end with one hand and let one lock at a time fall down, brushing it out from the roots with the body drush. Separate the hair on the dock and get right down to the skin to clean out all grease and dirt.

You can give a final polish over with the rubber, and lay the mane and tail by dipping the ends of the water brush bristles into water (*not* the horse's drinking water) and shaking off the surplus with a firm, sharp downward shake. Then brush the mane over from the crest and the dock hair on the tail.

Sponging should be done daily after body brushing, with the sponges damp but not wet; in winter the skin should be dried with an old towel.

Clipping

If a horse is to be doing anything more than light work (*ie* not much fast work, but maybe prolonged periods of trotting) in winter, you will have to clip off at least some of his long thick winter coat to avoid excessive sweating. Sweaty horses lose condition (weight) and energy and become chilled easily because although they may feel hot at first, they cool down more quickly than their coat dries off, and the combination of a wet coat and cold air makes them feel even colder than the daytime temperature; if neglected they could even suffer from hypothermia. It is possible to avoid this by making sure a sweated-up horse keeps on the move, walking smartly or trotting slowly, but if this is likely to happen regularly you might as well clip him at least partially.

For the one-horse owner, it is scarcely worth buying a clipping machine when so many professionals advertise regularly as being only too pleased to do it for you for a moderate fee. Look in your local horse journal or perhaps a farming paper with a horse section.

When presenting your horse for clipping he must be dry, and as clean as you can reasonably make him, because damp and dirt quickly blunt the clipper blades which then start to pull the hair and hurt him, making him understandably fractious. The machine will also heat up and the heat will probably bother him, too.

Some horses are difficult to clip because they don't like the noise of the machine or the vibration of the clipper head on their bodies, or the feel of the blades cutting the hair. However, it is an accepted fact, and has been scientifically proved, that if you feed a horse his favourite titbits or do something he particularly enjoys whilst the clipping operation is in progress, he will come to associate the task with something pleasurable and will co-operate much more. So while you are holding him, occasionally give him his favourite titbit, or at least have a net of good hay hanging up for him to nibble at.

As clipping proceeds, have a rug ready to put over his loins (the operator will work against the lie of the hair from back to front) to stop him feeling chilly and fidgety. Now and then, he or she will rest to give the horse a break and to clean and oil the blades. Have a radio playing soothing music, or stuff cotton wool into his ears, as this will help lessen the noise of the clippers – until you get to the head. This is the part many horses object to, and not surprisingly.

If it is absolutely essential for the horse's head to be clipped and he won't stand for electric clippers, try the small hand clippers used by hairdressers. And if all else fails, it's far simpler to get the vet to come and sedate him than have a fight. Some people do twitch horses (dealt with later in this section) and just for a short period such as the ears or part of the head it may be alright, but it's much better to achieve the job by means of persuasion if at all possible.

Heads can be neatened considerably by clipping or trimming with comb and scissors no more than the long hairs growing under the jaw, if you don't want to do the whole head.

WASHING YOUR HORSE

If the weather is mild, or you have access to a heated washing box, you may wish to shampoo a very dirty, greasy horse, and nowadays many people rinse off mud after exercise rather than using the old-fashioned practice of letting it dry on then brushing if off afterwards – the old theory being that rinsing off actually washed grit into the pores, causing mud fever! In practice, it is now felt by most people that brushing the dried mud off actually scratches the skin and causes irritation, and gives the mud fever bacteria easier entry to the skin, causing infection. The parts which get very muddy in winter are the legs, belly and breast (including in between the forelegs) and the sheath/udder area between the hind legs; here the skin is often thin and sensitive.

To shampoo a horse, use several buckets of clear, *warm* (not hot or cold) water and a mild animal or baby shampoo; I prefer to avoid the head. Soap the horse all over with a large car sponge, rinse very thoroughly indeed getting into all nooks and crannies such as under the jaws, between legs, behind pasterns and between the buttocks, scrape off excess water with a sweat scraper or the edge of your hand, and give a preliminary rub down with old terry towels. If the weather is warm, lead the horse round to dry. If it's chilly, put an anti-sweat rug on, or thatch him with straw, putting an old rug on top. Most modern rugs allow the horse to dry off because they let the moisture evaporate up through their fabric, but check this quality when you buy.

Stable bandages can be put on the legs, perhaps over sections of old mesh bed blanketing or coarse-knit dishcloths both of which allow air circulation. *It is essential that you dry the backs of the pasterns and the heels very well, perhaps using a hand-held hairdryer, to avoid chapped skin and cracked heels.* When using an electrical appliance in the yard, do so on a dry surface, ideally a rubber-floored grooming box, wear rubber boots yourself and connect the appliance through a standard circuit breaker available from any good electrical shop. Put excess flex in a non-metal bucket and ensure that the horse does not tread on it or interfere with plugs.

To wash the mane, simply soap it thoroughly right down to the roots, then squeeze clear water all down the crest with your large, well rinsed sponge. To do the tail, wash the dock as for the mane, and put the long tail hair into a bucket of soapy water and swish it around, rinsing it in clear water in the same way. It is most important to get all soap out of the mane and dock roots. Soak up excess water from the mane with old towels. For the tail grasp it at the end of the dock and, standing to one side with your back to the horse's head, try swinging the tail round hard (like a Catherine Wheel) and the water will fly out. Press it firmly with a dry towel and just let it dry.

If you return your horse to his box he'll probably roll to get rid of the uncomfortable feeling of being clean! If you turn him out he'll roll on the ground, though if there is a good covering of grass he won't get too dirty. If there is any mud around, of course, all your efforts will be wasted!

Opinions vary as to the temperature of water to be used; I feel it is best to use comfortably warm water for both washing and rinsing, as hot water could be uncomfortable and strange to the horse and will probably remove too much natural oil from skin and coat. If the horse is washed frequently this could lead to very dry skin prone to flaking, cracking and infection, not to mention a general tight and maybe itchy feeling which could cause the horse to feel uncomfortable and to rub.

Very many people happily use cold water, at least for rinsing off mud or soap, and particularly if using a hosepipe rather than buckets. I'm sure this must be as unpleasant for a horse, a warm-blooded animal like ourselves, as it would be for us, and he has no choice in the matter. Very reasonably priced instant water heaters are available (such as can be fitted to showers) and these are easily fitted to a hosepipe attachment, or the tack room sink from which you might run the hosepipe. The thermostat can be set at about 100.4°F/38°C which is the horse's body temperature, or a degree or two warmer, so you don't waste heat, and your problem is solved.

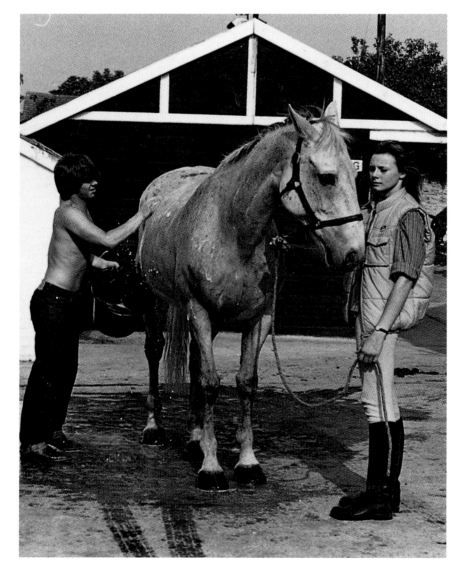

It's very refreshing to wash down a sweaty horse on a hot day. Plain tepid or lukewarm water is quite sufficient as too much soap, too often, removes too much natural oil and dries up the skin and coat.

An increasingly large number of people (myself included) feel it quite wrong and unfair to the horse to clip off the feeler whiskers which grow around the muzzle and eyes. These act like antennae and help the horse assess his surroundings, especially in the dark; they also help him to feed and generally act as an accessory to his sense of touch. Some horses become disorientated when these whiskers are clipped off, bang their heads, go off their feed and so on. They do come round eventually, but it is an unkind practice which I should like to see stopped.

Nor is it advisable to clip or trim out the hair inside the ears, as this acts as a filter for the dirt and bits which would otherwise fall into the ears and cause problems.

Give the horse a good body brushing after clipping to remove bits and 'hidden' dirt.

Trimming

Horses in their winter coats are not always clipped, particularly those with fine coats and/or in light work only, but they can be tidied up surprisingly well by judicious and careful trimming. In fact, trimming is needed all year round, and on clipped horses as well, to give that well groomed effect and make all the difference between a horse looking cared-for rather than slovenly. With a little practice you can do it yourself: you'll need a mane comb, an ordinary comb, a pair of curved, blunt-ended scissors and a pair of large, sharp scissors or shears; all your tools should be available from a tack shop.

Clipping is not a job for a novice, but you can help by soothing the horse and giving him tit-bits, also by helping smooth out 'wrinkly' areas. Here the handler is smoothing out the skin in the elbow area by holding the horse's leg forward. It is more comfortable for the horse to simply hold it out by the knee, and just as effective, the important point being to make sure the elbow comes forward.

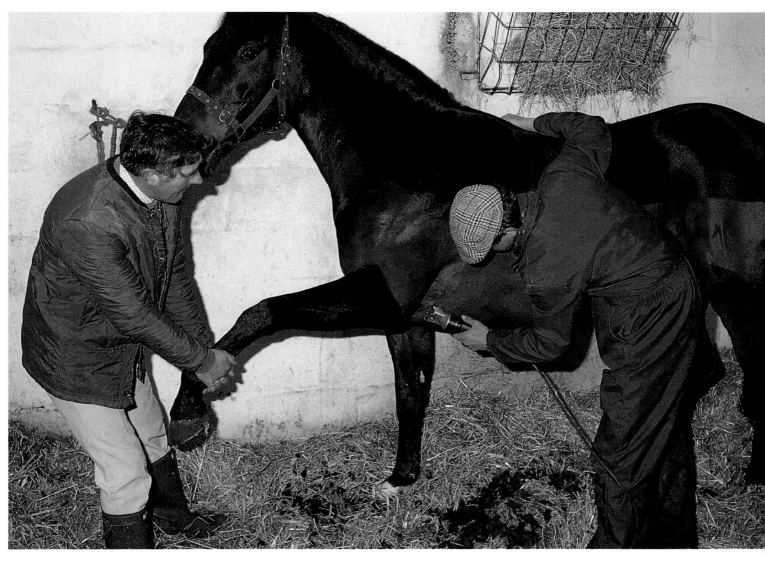

To pull (trim) the mane take your mane comb and backcomb the mane hair up to the roots. You hold the longer hairs in your left hand and comb the shorter hairs up them with the comb. At the roots, put the comb through the long hairs and hold them firmly, no more than six hairs at a time, against the teeth of the comb with your thumb, then snatch them out quickly. Do this gradually all along the mane, perhaps just a few hairs a day to stop the crest of neck getting sore and the horse objecting for ever more to this process. It is the longer, underneath hairs you pull out so the shorter ones are left, and fall smoothly over. If you pull out top hairs they'll look like a brush when they start to grow again.

To shorten a long mane that is already rather thin, simply snatch off the ends between thumb and finger at an angle, when they should break off quite easily. This method can be used on animals such as Arabs and native ponies which are not supposed to have their manes or tails artificially trimmed (or not so that it's obvious) in show classes.

If you feel this is beyond you, or you have a horse who simply won't stand for it, buy a razor comb. This has two rows of teeth with a blade set between, about half an inch down; you simply comb the hair at the roots in an up-and-under motion and the hair is magically cut off very near the roots underneath. Use the same comb on the ends of the mane to shorten a thin mane, if you wish, holding the hairs in one hand to make them taut and easily cut. With practice you get a very natural effect, and it can also be used on heels and fetlocks, up the backs of legs and under the jaw.

The more usual way to do these areas is to use your blunt-ended fetlock scissors and ordinary comb: comb the hair to be trimmed upwards, snipping off the hair which protrudes between the comb teeth with the scissors. On the fetlocks around the ergot (that little horny piece which is on the bottom point of the fetlock), leave a little tuft for drainage of water and to look more natural.

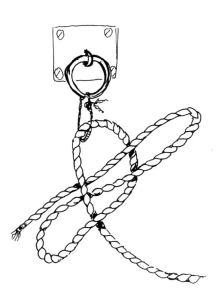

When tying up a horse, form an ordinary slip-knot. Most experts advise that the rope be tied through a loop of string which will break if the horse pulls back in panic for any reason, freeing him and avoiding a real fight in which he might lose his footing and injure both himself and you.

Some horses learn to pull the loose rope-end of a slip-knot and release themselves. If you pass the end through the slip-knot loop, like this, it prevents this, but if the horse then pulls the loose end he will tighten the knot and you cannot release him quickly in an emergency. For such horses, you can buy from some saddlers special quick-release clips which prevent the horse from fiddling but enable you to free him quickly, if necessary.

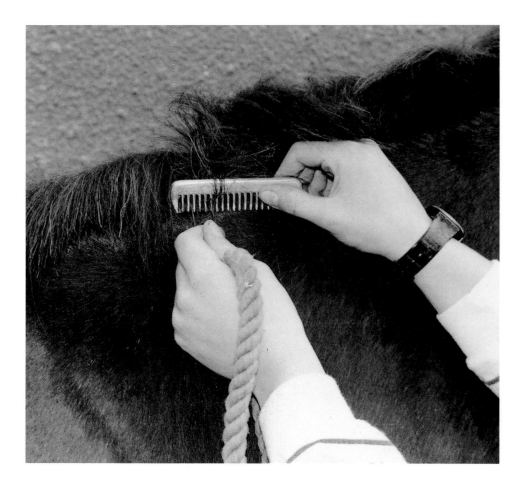

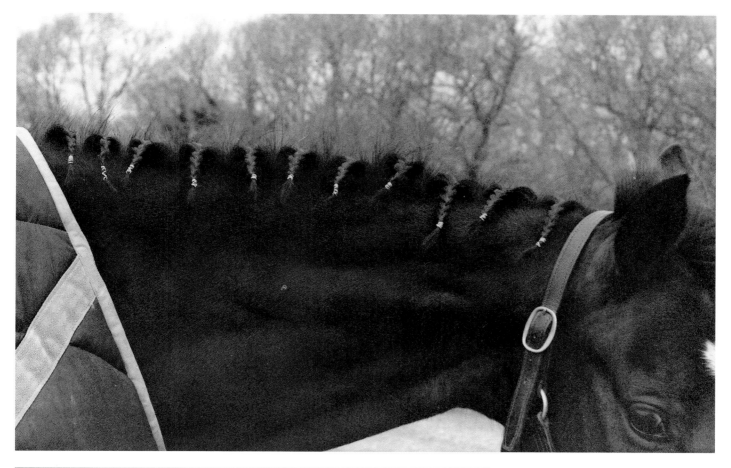

ABOVE:
To train a wilful mane to lie down, plait it into pigtails, like this, secured with elastic bands. There is no point in doing it tightly and it will be uncomfortable for the horse, anyway. The plaits can be left in indefinitely provided they are loose enough to be comfortable but most people remove them daily for grooming. You can damp the hair at the roots first and also use a little firm-hold hairspray down the crest after plaiting.

FAR LEFT:
To pull the mane to keep it a manageable length for plaiting, do a little each day to prevent soreness. Take a mane comb and hold the longest hairs at their ends, backcomb the rest up towards the crest then wrap the long hairs round the comb right at the roots, or hold them firmly against the comb with your thumb, and snatch them out quickly. Six hairs at a time is plenty.

LEFT:
To shorten a long mane which needs tidying but which cannot be pulled due to breed showing requirements (such as Arabs), take hold of a few hairs at the ends and pull them sharply to one side like this, snapping off the ends to give a natural effect.

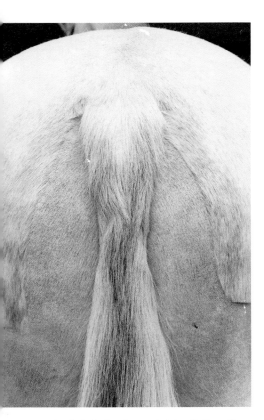

A pulled tail. The effect can be intensified by putting on a tail bandage, to keep the short hairs flat, for an hour or so after grooming.

RIGHT:
To pull the tail, take a very few side hairs on the dock and, as for the mane, wind them once round your mane comb or hold them with your thumb, snatching them out quickly at the roots, doing a little each day.

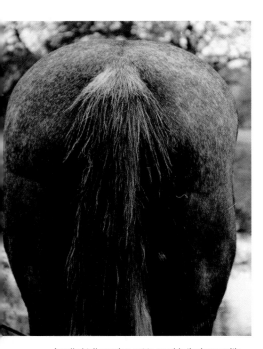

A pulled tail growing out to provide the horse with more protection for winter and to enable the owner to plait it up. It will take several months to grow out fully.

To pull and bang (trim) the tail you need to remove long hair from the sides of the dock, normally to about two-thirds of the way down. Just take two or three hairs, wind them round your finger, hold them firm with your thumb and snatch them out quickly, downwards. Start at the top and work down and, like the mane, just do a few hairs a day to prevent the dock getting sore.

To bang the end of the tail (cut it straight off across the bottom) ask a friend to put an arm under the tail to get it into the slightly elevated position it would be in when the horse is moving, hold the tail hairs in one hand and cut the ends off with your shears in one cut, sloping them *slightly* upwards towards the horse so that the end of the tail when free and in motion is dead level to the ground. Horses with a high proportion of Arab blood hold their tails higher than others, so for them you must make the slope a little more noticeable. Aim for having the bottom of the tail half-way between hocks and fetlocks when the horse is trotting, perhaps slightly shorter in winter so it doesn't get too muddy. Very long, or very short tails look equally silly.

Trimming the ears has to be done carefully to avoid hairs falling down inside them. With your blunt scissors, carefully close the edges of the ears together and snip off *only* the hair which protrudes beyond them, particularly the tufts at the base of the ears.

These techniques are quite enough to ensure you always have a horse which looks neat and well cared for, and which will be a credit to you and the horse world in general. Surely it's best to at least make sure you don't appear in public covered in mud and with bedding tangled in mane and tail!

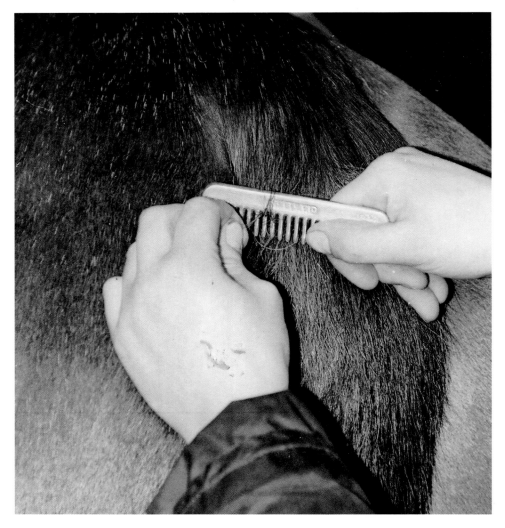

UNDERSTANDING THE HORSE

Many people are brought up with dogs in their home, and are quite used to handling them, training and feeding them, playing with them and generally associating with them. Very few civilisations have ever treated their horses in quite the same way, with the possible exception of the Arab Bedouin nomadic tribes, whose horses were quartered in their tents with them much of the time.

Many people who are familiar with dogs but new to horses tend, naturally enough, to treat them like very large dogs; they will go up and pat them on the head, whereas a horseman would know that it is better to stand slightly to one side and let the horse smell you first, then quietly raise a hand and stroke the horse on the neck. Dogs and horses could not, in fact, have a more different mentality; true, they are both gregarious animals living with others and in a strictly defined pack or herd hierarchy, but there the similarity ends. For in the wild, dogs actually prey on horses – the dog is a hunter, the horse its prey.

In their natural state horses live in herds usually made up of a single family group. There may be anything from two to about a dozen animals in a herd, which would comprise adult and young females, foals of both sexes, immature males and one adult male, the stallion. Males are kicked out of the herd when they become sexually mature and a threat to their father, the stallion. Females tend to stay with the family, and form particularly close bonds with each other – horse society is in fact matriarchal, with the real leader of the herd a dominant mare. She is the one who decides when to move on to fresh grazing and where to go, for example, while the stallion is mainly responsible for fighting off marauding stallions who might steal his harem; and when the herd is on the move it is the lead mare who will do just that, lead them, and the stallion will make up the rear, herding stragglers along with a low head and snaking movements of his neck.

Shying is the horse's natural reaction to what he sees as dangerous or alarming, the start of his instinctive flight-or-fight reaction. However, it can be unnerving to ride a horse who frequently shies, and can easily unseat you if your seat is not very secure. With increased trust in and respect for the rider, more schooling and a firmer seat it can be controlled. Shying is encouraged by too much food (energy-giving concentrates) and not enough exercise, when horses will shy for fun.

NOVICE RIDERS

Novice riders should never be placed in a position where they are allowed to handle suspect horses, for not only may they spoil the horse but they may be put in danger themselves. Professional instructors should be able to spot when a horse and rider are not matched, and should avoid putting them together until the rider has had more experience or the horse more schooling. If you intend to make a career with horses, there will come a time when you *have* to learn to handle more or less anything, although I have known several professionals who have had the great sense to admit when they just didn't gel with a horse and let someone else take it over. However, this is not weakness, nor is it 'giving up'.

EQUINE TEMPERAMENT

Most horses described as of dubious temperament or as downright nasty have been made so by ill-treatment, although there must be a few who are naturally so. In the course of several thousand years of selective breeding, man has bred out of the species most of the undesirable traits which made the horse difficult to handle and work with, and therefore of less use to him; however, now and then genuine 'bad 'uns' (to us) do appear, and these will rarely be suitable for a novice – they will always need expert handling, if they are to be tolerated at all.

One line of the horse business where temperament is put a definite second is racing: here, almost everything is sacrificed to the God of Speed. The requirement for speed and toughness has resulted in a population of horses with the constitution and conformation capable of withstanding the stresses placed on them in racing. And if a superlative racehorse or sire of winning stock has an 'iffy' temperament, his connections just put up with it as part of the package. This is not so in other spheres where most owners, particularly amateurs, expect their horses to be at least safe to handle even if they are not exactly affectionate. Mill Reef, however, was one example of a Classic winning racehorse who had one of the sweetest natures on record.

So, then, horses *are* individuals. One of the most basic tenets of learning about horses is to understand the oft-repeated saying that 'horses are individuals and must be treated as such'. This need not confuse you, however, particularly as the basic rules of general management such as feeding, bedding, exercise and so on are common to all horses.

The main characteristic shared by all horses is nervousness and being highly strung. This may sound strange when you think of quiet, plodding ponies, rock-steady police horses or calm, confident hunters, but underneath the surface all horses are unpredictable and easily startled, given certain circumstances. This is because for many millions of years they had to be constantly on the look-out for predators, constantly on the alert, ready to flee from every little rustle, strange sound, movement in the grass or shape behind a tree.

The horse's main defence is its speed, combined with a quick getaway. Its sprinting speed of roughly 35mph matches that of most of its natural predators, so if a horse allowed a predator – whether hunting dog, large cat such as lion, or whatever – to get really close, its chances of escape were no more than 50/50. Its advantage lay in spotting the predator while it was still some distance away – so while it was trying to hide and creep up to within springing distance, the horse was being careful to watch all around him minutely, and constantly flicking his ears for sounds, too. This naturally made for a nervous temperament.

Given a decent start, the horse's average survival ratio (though depending on the predator) in a chase is increased from one-to-one to four-to-one: for every four horses a predator attempts to catch and kill, it manages it only once. And then Nature stepped in with yet another advantage for the horse: it has far more stamina than any of its predators (one reason the horse is so useful to us). In a chase matched for speed, the horse can keep on running when the predator has run out of steam (although anyone riding a horse who has ever been chased by a dog in an open space such as a beach may doubt this!) So the horse relies on its alertness and speed but also its stamina and natural inclination to keep on galloping – often seemingly blindly – when frightened.

You'll have heard of the flight-or-fight response to threats. Horses would much rather flee from danger than stand and fight, unless they are cornered, and in a fight a man is no match for a horse's teeth and hooves, not to mention the half ton weight of the average riding horse.

Armed with this knowledge, we can begin to work out how best to handle the horse. We must also appreciate that most horses are followers, not leaders, and although they are intelligent in their own way and for their own natural lifestyle, they do not have the same *kind* of intellect as humans; furthermore they invariably have an amazing and unaccountable willingness to adapt to our society and to do what we ask. So we have an exceptionally useful and valuable asset on our hands – the horse; and the horse has been more useful to the human race and in more ways than any other animal on earth: no other animal combines its strength, speed and patience, its stamina, docility and trainability, and its sheer willingness to co-operate when it understands what we want and is treated justly.

I am not trying to wax poetic, merely to detail the qualities we have at our disposal and to provide an understanding of how to tap them.

Common to all horses but perhaps in differing degree, are the key watchwords to bear in mind when dealing with horses: quietness, calmness, confidence and, when called for, understanding firmness. If you are soft and weak with a horse needing leadership he could become a nervous wreck; if you are bombastic or rough with a natural leader horse, you really have to know what you are doing in order to win the natural dominance battle and avoid getting into a physical fight which you cannot possibly win.

The horse's instinct, especially in times of stress or pressure, is to behave as nature has been teaching him for millions of years. Despite all our training and influence even domesticated horses and ponies will almost instantly revert to nature when they are frightened, unhappy or in some way having a hard time. The flight-or-fight instinct is particularly strong, and horses see an unpleasant situation in the same way as they would see a predator. If they perceive pain, fear or something they cannot bear, they react instinctively by trying to get away. For example horses in pain, perhaps from a roughly used bit and heavy handed rider,

may pull harder and harder against the pain and may resort to bolting (running away) as the only way they know to escape.

On a practical basis, the horse's six main requirements for a contented life are food, water, shelter, company, space and movement. The first three have already been dealt with so far in this book, so let's look at the remaining three now.

Horses are herd animals and so almost without exception will only thrive when kept in the company of others. There is good sense in living in a herd community when you are a prey animal: quite simply, if a predator is on the prowl it knows that one horse/donkey/zebra carcass can feed its family or pack for several days, if necessary – certainly that particular predator is most unlikely to kill two large animals in one day. In a herd there is less chance of the predator picking *you* out of the crowd, and your hide may be saved to run away another day; and as long as you are mature but not old, and are healthy and sound, you stand an excellent chance of surviving. It is the very young and very old, the sick or injured, who are most vulnerable.

Horses, then, nearly always feel more secure when kept with others, stabled near them and turned out with them; certainly one other companion should be provided, and preferably several. Many owners *do* keep their horses alone, and the horses seem alright; but when taken into company it is my experience that they nearly all show their preference for being with other horses. It's their instinct. True loners are extremely rare in the horse family.

Horses kept alone can suffer from insecurity and loneliness, even at a low level which their owners may not detect, and will often form a friendship with another species such as a dog, a cat or a hen, even a cow or sheep. Humans are not usually available, since most owners spend no more than three or fours hours a day with their horses; and so we really cannot expect our horses to become as attached to us as do our dogs, even though they may give a good imitation of it at feed times or when they want to be brought in out of extreme weather.

A solitary horse is much more likely to jump out of its field, especially when it sees other animals in the distance or other horses going by; and then it can become a serious nuisance in the neighbourhood, quite apart from risking injury to itself as well. And it may become difficult to manage in company, too. My personal feeling is that it is unkind to keep a horse alone unless it has very regular contact (daily, if at all possible) with others, preferably in a situation where it can socialise naturally with them, such as when turned out in a paddock.

Personal space is something we all hear a good deal of these days, and horses have very definite requirements for this. It doesn't only mean that they need roomy

Horses rest and sleep in various ways. The horse in the background is grazing but on 'guard duty' at the same time, allowing others to rest. The horse on the right is dozing standing up and is able to set off in an instant should danger threaten. The horse in the middle foreground is resting on his brisket with his muzzle propped on the ground to take the weight of his heavy head. In this position he can experience short-wave sleep. The horse on the left is experiencing 'rapid-eye-movement' (REM) sleep, lying flat out in the only possible position for this sort of deep sleep.

enough stables (a subject already discussed), but like most of us, they need to maintain a private distance around them – this seems to be about four yards or so, and with some, a little more. If you observe a group of horses grazing in a paddock, you'll notice that only close friends within the herd graze closer to each other than this; there is always a discreet distance between the others.

If a strange horse or an unfriendly one comes any closer, the one approached may adopt his or her 'on guard' pose – standing to attention, ears pricked towards the intruder, nostrils flared to smell him, head and tail up to make himself appear bigger and more formidable should the intruder be thinking of attacking. This shows quite clearly that he resents his personal space being violated and that the intruder had better not come any closer. However, if the approached horse is lower down the herd hierarchy than the intruder, he may, instead of putting on a show like this, simply move off (fleeing instead of fighting) out of harm's way.

Owners should always take the trouble to observe the herd their horse lives with, both loose in the field and in the stable yard, and also out hacking. It is natural for some horses to be high up the pecking order and others to be low down, and you will never change this without a lot of trouble. So if a horse is seen to be constantly chivvied and bullied you must accept that he will never thrive in that environment; he should be put in another field, if possible with the horse next to him in the hierarchy, or at least with some other friendly horse or horses – it is up to you to allow him to live the contented life that in the wild he could ensure for himself, free to make his own choices and not a prisoner in a domestic field or paddock.

Do not stable your horse next to, near or preferably even within sight of an enemy. Horses in neighbouring stables are perforce very close to each other, often nearer than personal space dictates, and are separated only by the dividing wall – but they each know the other is there, and will never settle if they do not like each other. On the other hand, if friends are next to each other they will be the picture of contentment. If your horse is at livery and you feel he does not like his neighbour, you are quite entitled to ask for them to be separated. Any conscientious proprietor will understand this; indeed, in extreme cases, it could avoid destructive and injurious kicking matches between them which might result in a badly damaged wall and one or two lame horses.

Space is also preferred by horses in their grazing environment. It has been found that horses are happiest in domestic paddocks of at least ten acres or so (about five hectares), though this is quite beyond the comparative postage stamps available to many owners. There is a solution, though, if you can arrange it: make a regular habit of taking your horse on long rides in open country with other horses. This most nearly mimics the lifestyle he would have in nature and will go a long way to settling him and making him mentally satisfied and content. Picnic rides, bridleway discovery rides, competitive trail rides if you like the sport of long distance riding, and of course hunting: all these fulfil this need in the horse, so see what you can arrange.

And so to the final topic of the six – movement. We have seen how the horses which lead the life Nature intended for them roam in herds and are on the move for most of their 24 hours. It has been estimated that on average they spend about 16 hours a day in gentle exercise, usually walking about grazing, with the odd canter and play; they trot mostly when going some distance from one grazing ground to another. All in all they are programmed to move, and move they do. Imagine, then, the plight of a stabled horse given only two hours exercise a day. Even if that work is hard and tiring (which is unnatural, especially when combined with a further 22 hours of inactivity) this does not make for mental contentment or physical wellbeing – an amazing proposition when you consider that most of our top competition horses and racehorses are kept in exactly this way. I wonder how much better they would do (and feel) if they were also allowed out to play – as they are, in fact, in some establishments.

Many horses simply cannot stand this kind of lifestyle and develop aberrant behaviours (abnormal activities) to compensate for the distress they feel: they crib-bite, wind-suck and box-walk (these have already been discussed), all of which involve frequent movement and physical activity. Some kick their stables, either incessantly or in bouts; some take to kicking and biting people, and become impossible to handle when they are out, rearing, bucking, bolting, plunging, shying and generally making themselves very unsafe to ride, even for an expert. If they indulge in this kind of release behaviour in traffic you and the horse, other road users and pedestrians could all be seriously injured or killed – and all for the want of a little more natural freedom and movement.

It may be difficult, but do your absolute utmost not to keep your horse at any establishment which does not have year-round turn-out facilities, as many do, in winter particularly. The usual excuse at this time of year is that either the land is too muddy or the horse will ruin the land – but if you keep horses you have to expect this sort of thing, and must set aside some 'throw-away' area for winter exercise. Even if the horses are only turned out in pairs on a rota basis for an hour a day in addition to their exercise, it is better than nothing – and after an hour pestered by summer flies (because you forgot to apply the fly repellent!) or battered by winter weather they may be jolly glad to come in again!

This need for space and movement is my main reason for being an advocate of keeping horses yarded or on the combined system, or at grass with proper shelter. You just don't get the above problems unless a horse is already really confirmed in his vice – and then there is almost nothing you can do to stop him altogether, though most horses with vices do lessen their activities when kept more naturally. This subject is very fully discussed in my book *Behaviour Problems in Horses* (David & Charles) and there is not space to talk about it here; however, I hope this short discussion has warned novice horsemasters of the problems they can build up for themselves and their horses by over-restricting them in an unnatural way.

Horses are definitely more content when allowed out a good deal with others in roomy, well-sheltered fields like this one. However, over-stocking (putting on too many horses for the size) is bad for the land and horses if it is a permanent state of affairs. Land must be frequently rested and treated, under expert advice, if it is to remain in good heart (condition) and productive. These mares and foals will have sorted out their own herd hierarchy (pecking order) and the foals will become mentally well adjusted by associating with each other and with mares other than their dams. Playing with each other will also help them develop physically.

HANDLING HORSES

Probably one of the biggest aids you have in handling your horse is your voice. Horses have very acute and sensitive hearing and can, in fact, hear sounds at frequencies above and below those of our own hearing. They are sensitive to the tone of your voice as well as the words you use – which of course to the horse are simply sounds. They are also able to distinguish regional accents and can become confused by the same word said with a different accent, but they do learn a new 'language' (range of sounds), quite quickly.

Three examples of the above are known to me personally: a Thoroughbred stallion who, as a racehorse in training, was based in the north-east in a yard where he had obviously had a rough time as he now attempts to savage anyone with a Geordie accent. I also know a Morgan stallion who, when he was first imported from America, would not do a thing for his new owner until, fooling around one day, she started using an American accent, whereupon the horse was magically transformed and became instantly co-operative with, the owner reported, an immensely relieved look on his face! And the third example is that of a team of top Hungarian driving horses bought by a British competitor: within a week he had taught them, from scratch, to obey English commands (essential in driving horses who work mainly from the voice).

Much more emphasis could be laid on teaching horses and ponies to come to their names. This is basic training for dogs, yet no one bothers very much if a horse does not come when he is called; yet it can be so very convenient, and is a most useful way of gaining the attention of a distracted horse or one on the verge of fear or panic. Say the horse's name every time you approach him, before every request or command, and he'll soon catch on if he doesn't know his name already. When you visit him in the field call his name and *always* have a titbit for him; he'll soon associate his name with something pleasant. Always give titbits on the flat of your hand, so the horse doesn't confuse your fingers or thumb with the food. A horse can easily bite off your finger with no effort at all.

Horses respond best to simple sounds of one or two syllables. They are backed and trained (an expression commonly called 'breaking in' which I try to avoid) by use of these words on the lunge, taught to 'walk on', 'trot', 'canter', 'whoa', and by some trainers 'back' and 'come here'. You can continue to use these words during schooling and riding especially if you are having problems with transitions, but there are other words which are very helpful in daily handling in the stable, field and yard. One is 'over', to get the horse to move in the box when, for example, you are mucking or skipping out.

Place your hand confidently on the horse's side where your leg goes when riding so he'll link the action in his mind, push, and *simultaneously* say 'over'. If he doesn't oblige, repeat it exactly, saying nothing but 'over' or you'll confuse him and he'll never learn. If he still doesn't move, get someone to help you push and, to keep his balance, he'll probably move a foot just slightly. *Instantly* say: 'Good boy' in a very pleased tone, stroking him as well. Soon the penny will drop and all you'll have to do is put your hand on his side and maybe not even have to say the command, and he'll move. Once he cottons on to your routine and jobs, he'll watch you in the stable and will possibly even start moving over when you're ready but before you ask.

Teaching commands in this way will really help you when handling horses. The important thing to remember when training is that you must praise them instantly when they oblige, otherwise they will not associate the praise with what they have just done – if you wait one second, you have left it too late. It is the same with punishment. You have say a cross 'no' *while* the horse is doing wrong or he'll never learn it *is* wrong. Riders who are seen thrashing a horse several seconds after some misbehaviour are being extremely stupid on two counts. For one thing, a single stroke of the whip is quite enough to convey to the horse that the rider or

A mild form of restaint for a slightly difficult horse is to put the leadrope round the muzzle like this, but not so low as to hamper the breathing which can panic some horses and make them even more difficult.

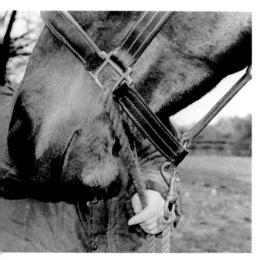

Pass the leadrope through both side dees of the headcollar and hold both ends together in your hand. This provides tension all round the lower part of the horse's head and gives more control, if mild, with a horse being slightly difficult. Note that the clip is fastened with the opening away from the head, which is both more secure and safer.

trainer is displeased; and secondly, as explained, the horse cannot possibly associate the thrashing with whatever he did to displease his rider if it is administered long after the event.

The use of the whip is a moot point anyway, and experts will argue over it forever. It is not necessary or desirable to shout at the horse, whatever he has done; but a firmly growled 'no', delivered the *instant* he does something you don't want him to do, tells him in no uncertain terms, and without the need for physical punishment, that such behaviour is not wanted. Only with stubborn horses who are repeatedly doing wrong when they *know* they are doing so, should the whip be used.

Safety

Many of the routine tasks you need to do – leading, picking up feet and so on – are shown in the accompanying photographs and captions, for clarity. The reason that correct ways have been formulated over the years for these tasks is partly because these ways work, and partly because they have turned out to be the safest ways. Horses are big strong animals and easily panicked, and anything we can do to reduce risks and increase safety should be done.

When working around horses it is best always to wear strong boots or shoes which will give your feet some protection should you get trodden on. Rubber boots and trainers, for example, give no real protection at all and you *could* get a broken or badly bruised foot.

When leading a horse who is at all difficult, you should wear a properly fastened hard hat and sturdy gloves, the hat to protect your head from the ground or his hooves should you fall or be pulled or pushed over, and the gloves to give you a firmer grip on the leadrope and prevent your hand being burned by friction should the rope be pulled through them. You'll also have more security if you tie a knot in the end of the rope.

When leading a horse on the road, say to and from his field, you should use a bridle with the reins brought over the horse's head; again, wear your hat, gloves and strong boots, and if possible take someone with you to help in case of difficulties. The last thing you want is a horse free on a traffic-bearing road.

Always walk on the horse's off (right) side and on the left side of the road, so you are going with the flow of any traffic and are between him and the traffic. It's surprising how many motorists think it's funny to frighten a horse and get too close to it, maybe sounding their horn at the same time; but if their first encounter is with a person rather than a horse, many might think twice!

Carry a long schooling whip in your right hand so that, used behind your back, you'll have fair control of the horse's quarters and can help prevent him swinging his quarters out into the traffic.

It's also a good plan to wear a reflective tabard over your clothing, even if it isn't dark.

Around the stable yard, be constantly on the lookout for anything that could spell trouble for the horse: grooming kit left lying on the floor or in the bedding; mucking out tools left carelessly on the ground; loose or even trailing bandages, and slipped rugs on which the horse might trip; feed and water containers out of their holders and left around for the horse to trip over or lie on; polythene bags, which are extremely slippery when trodden on; protruding nails in woodwork; broken windows; splintered wood; loose bolts or door hinges; cars parked carelessly – the list is endless, but it's up to you to cultivate an eagle eye for anything on which the horse could possibly hurt himself.

As for riding out, you can't do better than to take the British Horse Society's Riding and Road Safety Test, which will teach and prepare you how to use the roads in the correct and safest way. The society produces regularly updated literature on the test and on road safety in general.

Always use an approved hard hat, and a three- or four-point harness approved

LEADING HORSES ON THE ROAD

When leading a horse on a road, keep him well to the left (although not on the pavement of a public highway as this is illegal), walk at his right shoulder and lead from the bridle reins brought over his head. A bridle gives much more control than a headcollar or even a lungeing cavesson. Carry a long schooling whip in your outside (right) hand to help keep the quarters from swinging into the traffic. You can use it to flick the horse on the flank behind your back without turning round, altering your grip on the reins or losing any control. A hard hat, gloves and strong shoes or boots should be worn. If leading in dusk or dark conditions, the horse should wear reflective bandages and the tail be cut short enough not to obscure them. The leader should wear a reflective tabard and preferably carry a light. Cyclists illuminated cross belts, battery operated, are excellent as they leave your hands completely free and do not encumber you in any way.

THE TWITCH

A twitch is a wooden stick with a rope loop on the end. The loop is *carefully* tightened around the top lip so it is snug but not painful and is thought to stimulate, by means of accupressure, the body's own tranquillizing pain killers, so calming a horse down for unpleasant procedures.

CATCHING TIP

When approaching a loose horse to catch him, approach from the side and aim for the shoulder. If you startle him by approaching from behind where he cannot see you he could kick you.

Ready for a hack. The rider is, of course, wearing her hard hat, gloves and a reflective tabard even in daylight. The horse wears speedicut boots and knee pads.

TURNING OUT YOUR HORSE

When turning out a horse or pony when others are milling around, open the gate just wide enough for you both to get through and fill the gap without allowing others to escape. Keep hold of the gate with your free hand and hold up any surplus lead rope in the hand holding the pony.

Shoo away any other animals to a safe distance and come in quickly, closing the gate behind you and fastening it to prevent it swinging open.

Lead the pony a few paces into the field and turn her round to face the gate. Do not let her go immediately as she may get into the habit of charging off and dragging you with her, or trampling you. Stand at her head and remove the leadrope or headcollar. She will then wheel off towards the other ponies on her hindquarters and you will be well out of reach of her heels should she kick up in glee.

Always wear a properly-fitted hard hat with a three- or four-point harness, securely fastened.

to whatever British Standard is current; it will bear the usual kite mark of the British Standards Institue. Also advised is a back protector and, of course, sensible boots or footwear – *not* soft shoes or wellingtons.

Never, ever ride out in fog, and don't go out in the dark or dusk unless you have no choice. Always wear at least a stirrup light fixed to your right stirrup, and the palest clothing possible; horse riders usually seem to wear dull colours, and these are not the most easily visible to motorists, even in daylight. Wear reflective tabards, hat strips, bridle strips and bandages on the horse, and remember to keep the horse's tail short enough so his reflective bandages are not obscured. Reflective exercise sheets are also available.

When leading a horse to or from a field in dusk or dark conditions, carry a lantern, preferably showing white to the front and red to the rear (as do stirrup lights), and hold it in your right hand so that it can be seen. It is easier to handle than a torch and could be useful in an emergency.

Many people advocate riding a horse out in knee pads to protect his knees should he come down on the road. Most riders have to ride on roads for at least part of their ride, and it's an excellent idea to get your farrier to fit your horse with tiny road studs in all hind heels to give him extra grip. Alternatively, a special hard material such as borium or graphite can be welded into the heels and toes of the shoes; not only do these give grip themselves, but they also help prevent excessive wear of the shoes.

Good manners both of horse and rider not only make life more pleasant, but easier and safer. Horses should be well trained and responsive to commands, and disciplined without being browbeaten. On the roads, be considerate to other road users, particularly as most drivers do not understand horses and many seem not to care or even to resent horses using the roads. Nothing is gained by having rows. If you genuinely feel you have been a victim of dangerous or careless driving, try, if at all possible, to remember the number and type of car and report it to the police. (By the way, it is not considered safe to ride with a pen or pencil in your pocket in case it sticks into you in a fall.) Be sure you act correctly at all times, and do not venture out with a horse known to be unreliable in traffic; they all have to be trained, but that is no job for a novice.

It's safer to ride out in company, but in any case always have on you either inside your hat or in a pocket, your name, address and a telephone number where help can be reached, or contact made, should you end up unconscious in a fall or accident. To your saddle and/or bridle, fix dog tags bearing the phone number of your yard and perhaps the address, should the horse get free and be found wandering miles away. It has happened.

Riding *is* a risk sport, but by following sensible guidelines the chances are that you will never have a serious accident.

FARRIERY AND THE HORSE'S FEET

The foot is a highly complex, finely tuned structure of bones, sensitive fleshy tissue and horn, with a generous supply of blood and nerves.

When the horse is standing normally the part you see is the wall which is made of horn – this is related to skin and hair, and consists largely of a hardened protein called keratin. The wall grows down from the ridge round the top of the hoof called the coronet, equivalent to our cuticle; inside, the coronet is richly supplied with blood which carries the nutriments needed to form the horn. If the diet is poor, so will be the quality of the horn; it may become weak and brittle, and will tend to crack and chip or flake away, possibly exposing the sensitive structures, thereby laming the horse and allowing infection to occur. If the coronet is injured by a knock or tread, the horn growing from it could be damaged, temporarily or permanently; damage such as this may grow out, or the horse could be left with a permanent ridge, groove or crack in his hoof – and it can take six months or more for a horse to grow a complete new hoof.

The underside of the hoof consists of the bottom rim of the wall, the slightly arched horny sole and the frog of rubbery horn. There are grooves at the sides and in the middle of the frog which provide an uneven surface to help the horse get a purchase on the ground. The two fleshy bulbs at the back of the hoof are the heels.

The wall rim on the ground is known as the bearing surface because it bears on the ground (or shoe) and carries the animal's weight.

Inside the hoof the inner surface of the wall is lined with insensitive leaves of horn rather like the underside of a mushroom, and these are called the horny laminae. Interlocking with these are sensitive leaves called (predictably enough!) the sensitive laminae. When a horse's feet are trimmed by the farrier, you will notice an area of whitish horn called quite simply the white line, which indicates where the insensitive and sensitive laminae meet inside the foot. A farrier can nail into the rim or just into the white line safely, but if he goes inside the white line he will hit sensitive tissue and lame the horse.

The sensitive laminae are bonded to the outside of the pedal bone which gives the hoof its crescent shape. Behind this is the navicular bone (formerly called the coffin bone because of its shape), and resting on top of the joint between them is the short pastern bone, lying half in the foot and half out. Around this is the

The horse family is the only one in the world with a single-digit foot, and as such is regarded by zoologists as the world's most specialised running animal. Because its long legs have no muscle below the knee and hock, and with its toe, the hoof, it can run at considerable speed (although it is not the fastest animal in the world) and does not fatigue easily, unlike its predators. The hoof of today's horse is actually the last remaining of several toes: the earliest recognisable horse ancestor, 50 million years ago or so, had four toes on the front feet and three behind, rather like the present-day tapir, a relative of the horse.

As the horse evolved and required more and more speed, the other toes gradually dwindled away – the one remaining is, in fact, equivalent to the tip of our middle finger or toe.

coronet. Above the short pastern bone is the long pastern bone, and it is this one which to the outside eye forms the pastern. The cannon ('shin') bone joins with the long pastern bone to form the fetlock joint, together with two small bones at each side called the sesamoid bones.

Lining the inside of the horny sole and frog are the sensitive sole and frog, and immediately above the heels inside the foot is a firm spongy wedge of tissue called the plantar or digital cushion.

In addition to these main structures there are ligaments and tendons which respectively bind the bones together and provide attachment for the muscles up in the forearm, from which they run. The tendons are attached to the outside and underneath of the pedal bone, and when the muscles of the forearm (or thigh) contract, they pull on the tendons forcing them to flex or extend the leg. The ligaments help to prevent too much movement, and provide stability. The joints between the bones are miracles of engineering and chemistry: the ends of the bones are protected with gristle (cartilage, like the end of our noses) and a special lubricating fluid called synovia, the whole being encased and sealed in a capsule of tissue. The cartilage and synovia both help cushion the joints and protect the bones from undue wear.

Parts of the foot:

A long pastern bone
B digital extensor tendon
C coronet/coronary band
D hoof wall
E pedal bone
F sensitive sole
G insensitive sole
H superficial sesamoid ligaments
I digital flexor tendon
J short pastern bone
K suspensory ligament of navicular bone
L navicular bone
M digital or plantar cushion
N sensitive frog
O insensitive frog

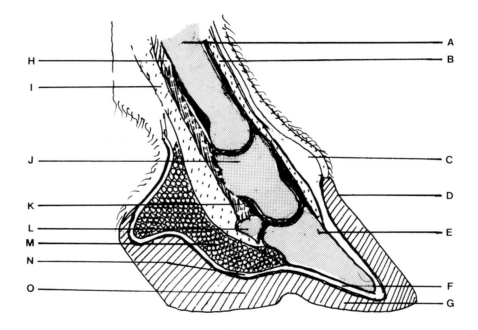

It used to be thought until very recently that the frog was the main anti-jar mechanism of the foot, as well as helping to provide grip. Much was said about the 'frog pressure theory' which held that without pressure of the frog on the ground during movement the blood could not be pumped back up the leg, making way for new nutrient- and oxygen-rich blood. However, current research is showing that although the frog helps, the main blood-pumping mechanism is, in fact, the flattening and slight expansion of the whole foot, particularly at the heels and quarters (sides), when it bears the bodyweight; the pressure squashes the soft, sensitive tissues inside the foot (including the plantar cushion), forcing out the used blood (venous blood in the veins) which is pushed up the leg into the veins on its way back to the heart.

The veins are equipped with valves which prevent the blood flowing back down the leg into the foot, so when the horse lifts its hoof, fresh arterial blood (in the arteries) – coming from the heart and carrying food and oxygen – surges into the foot. The veins and arteries mesh minutely in the foot in smaller and smaller blood vessels called capillaries making the changeover possible.

Need for shoes

In wild conditions, the feet do not normally overgrow unless an animal's movements are restricted because of injury or prolonged disease – in which case predators would not let it survive long, anyway. Varied terrain and the horse's inclination to exercise control the amount of growth, and a natural, varied diet would normally ensure that the hooves receive all the right nutrients to ensure horn health.

In domesticity horses work, usually on hard roads, and this wears down the horn faster than it can grow. Early civilisations soon realised what was happening and the horse shoe (initially boots of leather) gradually evolved. But keeping the shoes on all the time – as our modern, nailed-on shoes are – prevented any wear at all, so the feet then overgrew and needed regular trimming to maintain their correct shape and length.

Most riding horses in the western hemisphere wear a shoe based on the English hunter shoe and these are usually very similar in design, although there are specialised shoes for special purposes. Shoes are made of mild steel rather than iron, and most farriers buy in a stock of ready-made shoes in various sizes which they then alter on their forge to fit the horse's foot. A farrier is the correct word for someone who works with metal and shoes horses; blacksmiths work with metal, often doing intricate wrought iron work as well as general metal work, but they don't shoe horses. Hand-forged shoes certainly still exist but they take much more time to make and are beautiful examples of craftsmanship – as is a well-shod horse.

Whereas heavy draught horses would wear shoes of 'plain stamped' type (plain bars of shaped metal), hunters and riding animals need a much lighter shoe; theirs are normally shaped away on the inside edge (called 'concaved out') making it easier for the horse to pull his feet out of the mud without the shoe being sucked off the foot, and they are fullered (grooved) on the ground surface, partly to reduce weight and partly to give better grip on the ground. Holes can be made in the heels to take a variety of studs for different going (ground conditions) such as very hard, or deep mud, frost and so on.

There are three main weights of shoe: lightweight, which are suitable for unfit horses or those doing little roadwork; medium weight, suitable for most purposes; and heavyweight, suitable only really for heavyweight hunters and so on doing a lot of road work. Racehorses during a race will wear very light aluminium shoes called racing plates, and for exercise they often wear light (lighter than normal lightweight) steel shoes called race exercise plates. Show horses also sometimes wear this sort of plate, or even aluminium showing shoes.

General purpose shoes or those used for active work such as hacking, jumping, competitions and so on normally have a raised clip of metal turned up against the toe (a clip) to help keep them in place; on the hind shoes there are two clips, one on each quarter, so that should the horse be inclined to tread on his fore heels the injury may not be so bad as if he had a hind toe clip. Hind shoes often have the toe 'rolled', or hammered back a little, to help guard against this, too.

The shoeing process

The farrier should first see your horse walked and trotted up, even if he already knows him well, to check if there has been any change in his way of going which could indicate a developing problem or lameness. He may speak to the horse for a minute, perhaps give him a titbit or a bit of fuss, and will then remove his old shoes and study them for areas of wear which indicate how the horse puts weight on his feet and moves them. He will trim excess horn off the feet with pincers, drawing knife or rasp, correcting any serious misshape but keeping basically to the horse's natural foot shape and size.

The foot, if it is normal and healthy, should never be modified to fit the shoe – it is the shoe which should be made to fit the foot. Otherwise undue stresses would

CARE AND MANAGEMENT
Need for shoes
The shoeing process
Your farrier

YOUR FARRIER

Farriers are highly skilled men (rarely women) who follow a long training and apprenticeship, and have to be qualified and registered with the Farriers Registration Council. They may have qualifications such as AWCF (Associate of the Worshipful Company of Farriers; or their Fellowship, which is FWCF), and there are others such as AFCL, or DWCF; they may have an extra merit award – 'Hons' (with Honours)–after their letters. Ex-army farriers may have 'BII' after their names.

You can find a farrier by word-of-mouth recommendation from either your adviser, your livery stable or riding school proprietor, or maybe from other horse owners, or your vet may recommend one he or she works with sometimes; or you could contact the Farriers Registration Council at PO Box 49, East of England Showground, Peterborough, PE2 0GU to find out which farriers work in your area.

Most farriers are over-supplied with work and clients, and it is understandable if they refuse to do work for people who do not pay their bills on time (often straight after the job), who present horses with filthy legs and feet, who expect the farrier to go out and actually catch the horse up from the field, or who own horses which are difficult to shoe. This last can be a real problem and farriers are, in fact, trained to cope with difficult horses, but it is also true that some won't bother to deal with them. Your farrier is absolutely essential to your horsey operations so do all you can to treat him as well as you would your vet. Pay up promptly, show considerable interest in his work, discuss your horse with him and seek his opinion, and do all you can to help your horse overcome any fears by quiet, persistent treatment and rewards for being good.

A useful in-between shoeing method for horses who dislike being shod hot is to shape the shoe hot, cool it down (by plunging in water) before trying it on the foot, then re-heating it so that alterations can be more easily made. This produces a better fit than cold shoeing yet avoids the burning-on process which disturbs some horses because of the smell, smoke and hiss, and also avoids the possibility of excessive burning-on and drying out of the horn.

occur which could, in time, cause lameness either in the foot or further up in the leg.

Some yards have their own forge (fire) where the shoes can be heated up and altered, and many farriers travel with a portable one. Some owners take their horses to a forge (the building) – having been out of fashion for a while, this practice is now returning to favour with very busy farriers who don't have time to travel, and whose area is well supplied with clients.

There are two basic methods of shoeing, called hot and cold. In hot shoeing, the shoe is heated until it is red hot, then pressed very *lightly* against the bearing surface of the wall; the farrier holds it with a pritchel, a shaped rod pressed into one of the nail holes. If there is an even burn all round the bearing surface when the shoe is taken away, the shoe is pressing evenly on the foot, and little if anything will be needed in the way of alterations. If some areas are burned and some are not, this indicates that either the foot has been trimmed unevenly or the shoe is uneven, and the farrier will make corrections to foot or shoe accordingly, as uneven contact causes uneven stresses on the foot and leg which could well result in lameness. The shoe is cooled in water and then nailed on.

As it implies, in cold shoeing the metal is cold, and cannot be so easily altered (although changes are not impossible). However, if the farrier knows your horse and is good at his job, this method can be quite satisfactory – racehorses are usually cold shod (but then lightweight plates are easier to alter).

Horseshoe nails are specially shaped to curve slightly outwards when the farrier nails them in. Ideally they should all come out of the wall of the hoof at the same height unless there are difficulties with broken horn which the farrier has to avoid. There are normally four nails put in the outside of the foot and three on the inside, and the nails do not go right back to the heels because this is the area of the foot which needs to expand most to maintain blood circulation and foot health.

The nail ends are twisted off with the claw end of the hammer, and the metal left protruding, the clench, is turned downwards towards the shoe into a little groove the farrier has made with the edge of the rasp, or by means of a tool called a 'clencher'. This secures the shoe on the foot; if it were not for the clenches the shoe would be likely to pull off in mud or as the result of treading, or simply drop off.

The clips are tapped lightly into place against the wall, and some farriers cut little spaces out of the horn with the drawing knife for them to fit into. The clenches may be rasped smooth, although too much rasping of the horn is a bad thing as it predisposes to loss of moisture from the foot and can result in cracking horn and lost shoes. The edge of the rasp will be drawn round the join between shoe and foot to tidy it up, and finally hoof oil may be applied. And so the job is complete!

The frog and sole should, during trimming, have only surplus, loose flakes of horn removed with the drawing knife. The grooves and cleft of frog may be cleared of overgrown horn which is blocking them and could set up infection, but should not actually be cut away to 'open them up' as is occasionally done. Opinions vary on this topic and some farriers are very set in their view, taking questions and opinions from owners as criticism of their work! However, it is your horse and you are trying to learn, so discuss this with the farrier, or any topic you don't understand, and get his opinion. A good farrier will appreciate your interest and hopefully you can come to a mutually satisfactory understanding over points of difference.

In a well-shod foot, there should be no gaps between foot and shoe, the nails (clenches) should be as near the same height as possible, there should be no nails in the heel area unless unavoidable because of broken horn elsewhere, and the shoe should have been made to fit the foot, not vice versa.

After shoeing the horse should be trotted up on a loose rope on a hard surface to check he is sound and has not been accidentally pricked in the sensitive areas of the foot by a nail (unusual but not unknown). Sometimes the burning on process will

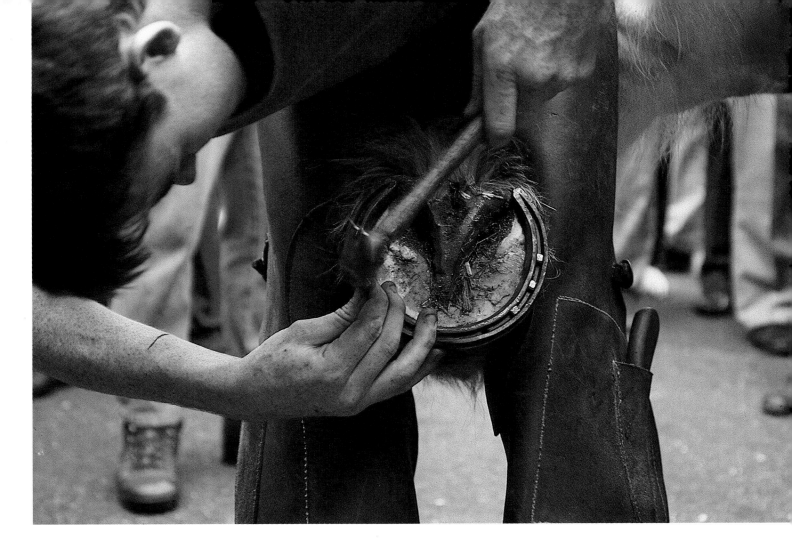

make a horse with sensitive feet go slightly short (lame), but this should pass within 24 hours. The burn should be light and short, as horn singes immediately. Such horses should be shod cold, as hot shoeing obviously causes them pain or discomfort.

Most horses need shoeing about every six weeks depending on their rate of horn growth and the amount of road work they do. However, don't make the mistake of putting heavy shoes on a horse to last longer – if they're too heavy they will be hard work and cause undue strain to lift, the horse will put his feet down that much harder which will concuss his feet and legs, and he will actually wear out the shoes more quickly. You can't win!

If shoes are not worn too badly by the time the feet are long enough to need trimming, your farrier can do a remove: that is, he removes the shoes, trims the feet, then replaces the same shoes.

In a very well shod foot the shoe may be worn razor thin yet still be firmly in place with no risen clenches. In a less well fitting shoe, the clenches will be pushing up out of their holes as the nail heads wear down and become loose in their holes, perhaps because they were the wrong size; the horn may be growing over the edge of the shoe, or the shoe may have slipped and be pressing into the heels of the foot.

This latter point is also important when looking to see if the foot is well shod. It is becoming quite common in some areas for the horse to be shod 'short' *ie* the shoe does not reach right back to support the heels – and over a period of a few months, this can result in weakened heels and lameness. The heels of the shoes should therefore fit well back so they protect the heels fully. To help prevent a horse treading on the heels, they can be more tapered (pencilled) and rolled under slightly.

This photograph, taken during a shoeing demonstration at the Irish National Stud, shows a shoe which has been well fitted right back to the heels to give essential support. Note that there are no nails in the heel area: this is to allow the foot to expand as much as possible in that area: this is to allow the foot to expand as much as possible in that area, an action which is important in ensuring a healthy blood circulation in the foot.

Risen clenches are particularly dangerous, as the horse can rip open his opposite leg when in action; the shoes may also come loose, making that characteristic clanking noise. Horses with very loose shoes should not be worked until the shoe has been taken off (in which case work on a soft surface) and/or the farrier has attended to it.

When the farrier has finished, make another appointment in four, six or eight weeks time, according to what your horse needs, so you can be sure of getting attention when needed. Don't make appointments and then not keep them – this is a sure way to alienate a farrier, vet or anyone else! If you are considerate, your farrier is much more likely to come out to you in an emergency such as a lost or twisted shoe.

Foot care

A very common complaint from farriers is that owners don't look at their horses' feet from one shoeing to the next. They seem to think farriers are miracle workers who at one shoeing can put right weeks or months of neglect resulting in cracked and broken horn – perhaps because shoes have been left on too long – bruised feet (including corns, which mostly occur in the heel area) and the development of thrush. All these could be prevented by a few minutes of routine daily care.

The feet should be picked out and the shoes checked before and after every ride. Anyway, it 'isn't done' to take the horse out with his feet packed with muck and trailing droppings and bedding all over the yard, and if he has a loose shoe you need to know about it beforehand. On return, again you need to know if he has loosened or spread (twisted) a shoe during work and must check that he hasn't picked up stones, bottle tops or any other debris in his feet which could cause pressure and eventual lameness if left wedged in the underside of his feet. He may have cut his heels or frog on glass or a sharp stone, may have picked up a nail – anything – so do check, particularly as he may not necessarily be lame immediately on return.

Horses coming in from the field in wet weather usually have their feet clogged with mud, unless the field has a good covering of grass. Some people purposely leave the mud in, maintaining it protects the horn from the manure that the feet will subsequently encounter in the stable – better for the feet to be clogged with mud than manure, they reason, and I see their point; but I still feel it's better to pick out the feet and examine them for injuries. The mud just might contain a stone which will cause lameness by next morning – besides which, if the stable is kept clean as it should be and the feet picked out regularly, manure will not be a problem.

If your horse is kept standing for many hours a day on filthy wet bedding, however, his feet may well develop thrush: this is a fungal disease affecting the frog, with a foul smell and dark discharge. It can also occur in horses turned out on constantly wet land, the softened horn making an easy access point for the infection.

Surgical shoeing

Your farrier is well qualified to work closely with your vet in cases of foot and leg disorders; by means of surgical shoes he can, in conjunction with the vet or otherwise, greatly relieve any pain and difficulty of movement in the horse, and assist his recovery. However, surgical shoeing is a great skill, as each case is different and shoes have to be very carefully made and fitted by hand.

In case of sprained tendons, for instance, the farrier might fit shoes with thickened heels to raise them and thereby take some of the stress off the tendons. Laminitis cases can be fitted with heart-bar shoes: these put carefully gauged pressure near the point (end) of the frog, which helps prevent the pedal bone inside the foot moving downwards and maybe piercing the sole, as it does in bad cases of the disease. Bad laminitis cases with heart-bar shoes can often manage to walk, where otherwise they would prefer to stand rooted to the spot with their weight on their heels to take the stress off the front of the feet which hurt most.

These are only two examples of surgical shoes. The whole subject is very complex and involves great skill, but basically the farrier can often, with appropriate shoeing, help your horse make a more comfortable and effective recovery than if he were simply left in normal shoes, or with no shoes.

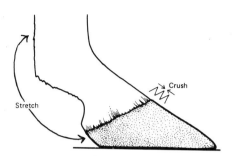

The balanced foot

Even trimming the foot calls for a good eye for proportion in relation to the horse's action. Ideally the foot should meet the criteria described in Part 2 if it is to be properly balanced and hit the ground evenly, sending even stresses up the leg with which it can cope. Lack of balance can result in one part of the hoof hitting the ground before the rest, wearing and stressing not only the hoof but, for a split second, putting undue force on that side of the leg as well, as the force travels up it. The hoof may then put out more and thicker horn on the over-stressed part of the foot, and in time the foot may become grossly misshapen; this will encourage lameness, cause difficulty in shoeing, and may even hamper the blood supply which can result in real troubles such as laminitis or navicular disease.

Diagram indicating the stress areas in a foot with a long toe and low heel.

Familiarity breeds contempt, so they say; and it is very easy with a horse you see every day not to notice that something is going a little wrong with a horse's feet. It pays, therefore, to make a point of taking a close look yourself at your horse's foot balance, at least when he is shod, so you can discuss any dubious points with your farrier. If, for example, when picking out his feet, you notice a particular shoe wearing unevenly, you should discuss it with the farrier when he comes: in the meantime watch to see if your horse's action is changing, if he is perhaps swinging a leg out of true when moving, or perhaps going unevenly or with shorter strides and less freely and willingly than usual. These are all signs that he is not feeling comfortable, and the chances are that the trouble is in his feet.

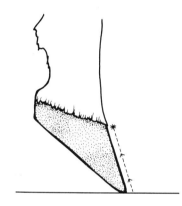

Corrective trimming

Corrective trimming is just that – correcting foot shape and balance by means of trimming. In a horse whose feet have been neglected and have therefore become misshapen, corrective trimming can gradually put him back to rights, but only if he had good, balanced feet beforehand. In fact, corrective trimming of this sort is something which has most relevance in foals as the bones are fairly malleable up to six months of age, and deformities or slight variations from true can be corrected quite well, the younger they are started the better being the results.

The above-shaped foot 'breaking over' in action, indicating the excess pressure directed up the toe to the coronet, resulting in a crushing of the horn-forming structures in the coronary corium.

If a foal, for instance, is tipping over a little on to the outside of one foot, the farrier will take more horn off the inside of that foot so that his weight is guided over to a more central position, counteracting the tendency to put too much weight on the outside of the foot, and, therefore, the leg. Many foals have what are termed 'weak heels': they do not have much height in the heels and may even appear to be going 'down on their fetlocks', with the fetlocks being too near the ground and the pasterns almost horizontal; this is more common in the hind feet than the fore. Sometimes this defect rights itself as the foal matures and grows, but if it appears that no progress is being made the farrier can, by corrective trimming and making tiny supportive shoes, put the foal on the road to correctly formed limbs and feet.

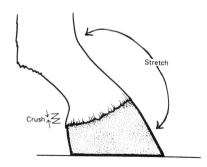

A hoof of the opposite confirmation to that shown in the illustration at the top of the page, showing the stress areas.

After the age of six months, however, if a horse or pony does not have naturally well balanced feet, there is little the farrier can do to change his natural conformation – the bones of the feet and legs are too hard to be greatly affected. In fact if too much is attempted, the horse can be made extremely uncomfortable because of his limbs being forced to bear unaccustomed stresses as the balance of his foot and therefore leg are changed. This can even lead to lameness. Slight corrections can certainly be made, but nothing drastic – that is, unless the lack of balance has been caused by neglect and not by natural development. This is yet another skilled and fascinating subject to discuss with your farrier, and even a busy farrier, if he is conscientious, will appreciate your interest.

CARE AND MANAGEMENT
Pads and studs
Exercise and fitness

Pads and studs

Brief mention has already been made of studs; there is a bewildering variety, but they each have a purpose: long pointed ones, for example, are for cross country work in slippery mud, square ones for hard ground, and small ones for slippery roads. Those with a roughened core of hardened material, which shows on the outside as a blob of different metal, are for icy conditions; and let it be said that although it is normally unwise to ride in icy conditions, some horses do play polo and race on ice in Scandinavian countries, in Austria, Canada and also in Switzerland, when the are fitted with special shoes, studs and frost nails.

The important thing to remember about studs is that if they are only fitted on the outside heel of a shoe, and normally just the back – as is so often advised – this is going to throw the foot out of balance; and if the studs are kept in when not particularly needed, the horse will feel as though he is stepping on a pebble all the time. Many horses (police, for instance) have studs in *both* hind heels of both hind shoes to even them up and give good grip. The reason some people disagree with this practice is that should the horse tread on himself it will be with the toe or heel of a foot, and a heel wearing a stud can inflict a much nastier injury than one with an ordinary shoe.

Both opinions have a point, but your farrier will help you make the right decision for your horse, if he needs studs at all. The things to remember are: never to leave studs in when they aren't needed (you can unscrew them and plug the holes with oiled cotton wool to keep the threads clean); and to use the right studs for the right conditions – then you won't go far wrong.

Pads are another area in which opinions vary. They are intended to be used on horses with tender feet to protect thin or low soles from bruising on rough going, or to lessen the jar caused by metal shoes hitting hard road. However, most healthy horses on good diets and with well cared for, well conformed feet do not need pads and their use is not widespread.

The problems lie partly with the materials from which they are made, normally polyurethane these days, but leather ones are still sometimes used. Both materials have some give in them and this, many people find, means that the constant slight movement of the pad, fixed as it is between shoe and foot for cushioning protection, can cause the shoe to move and loosen.

Some pads just fit between the shoe and the foot, but others extend across the whole of the underside of the foot, covering the sole and frog, and if the material is capable of absorbing and holding water, as are both polyurethane and leather, this can actually encourage thrush and softened horn to develop.

New materials are coming out all the time, however, and part of the research concerning the foot is looking into whether pads do actually lessen jar, depending on which material they are made from. For general purposes though, you are unlikely to need pads and should not fit them unless advised to do so by your vet or farrier – and even then they may not help or be successful.

Note the more oval shape of this hind foot (compare it with the fore foot on page 79). It has been fitted with just one stud in the outside heel and quite a large stud: this will seriously unbalance the foot and cause uneven stresses on foot and leg if the horse is worked on roads or hard ground. In softer ground the stud will sink in and help prevent slipping.

EXERCISE AND FITNESS

There is an old adage which says: 'slow work is always beneficial, fast work rarely is'. However, there are times when even slow work is dangerous, such as with a dead lame horse; and at times fast work is essential to develop wind and heart – but basically that old saying is true. You'll hardly ever do your horse any harm, and will probably do him a lot of good, if you make a practice of taking him on long, active walks, with nothing faster than a working trot on roads.

When I was small, my old ex-cavalry riding master used to say that we should 'walk fast, trot slow and canter in between', meaning that for general riding purposes the pony should walk out well up to his bridle, trot steadily to prevent jar on his precious legs on the roads, and canter on a suitably grassy or maybe sandy area in a nice, easy, in-between pace which was comfortable for the rider and easy for him to maintain in accordance with his natural rhythm – for like cars, each horse and pony has its own natural 'cruising speed' in all paces, but particularly canter. If you allow your horse to choose his speed for himself (provided he is not one of those who persistently lags and slouches along or, conversely, charges off like an old-fashioned steam train), your exercise will be beneficial to him and not injurious; and because he is easy and comfortable within himself, he will be comfortable and enjoyable to you, too.

This does not mean, of course, that you cannot enjoy a controlled gallop when conditions, and your horse's fitness, warrant it – that's part of the fun of riding – but it's the easy, regular exercise rather than the more stressful, energetic sort which benefits your horse and makes him fit enough to actually do the galloping and jumping which most of us find so exciting.

On any two-hour ride, for example, a horse should spend at least half of it walking, with most of the remaining hour spent in steady trots with perhaps a couple of canters. There is no need to be rigid, but this gives a fair idea of what a hack consisting of balanced exercise for a reasonably fit horse could be like. Galloping a horse into a sweat, unless it is being done for short spells as part of his fitness programme, is not beneficial, and too much of it can over-tire the horse and cause loss of condition.

It is also important to warm the horse up properly before faster work, by walking the first mile or so out. Muscles need to be put to work gradually to get an adequate blood flow coursing through them, bringing oxygen and energy and ready to take away the waste products produced by and in the muscles in the course of the horse's work. If the muscles are not adequately serviced in this way they cannot work properly and could suffer injury (be pulled or torn).

You should also walk the last mile home – another old and correct saying. The reason for walking the horse to cool him down is to keep the blood flowing at a moderate pace to ensure that the waste products *are* effectively removed after work, and that energy and oxygen are delivered in sufficient quantities to help the muscles recover quickly after their efforts; particularly as these waste products can be toxic, and if they remain too long in the muscles they could damage the cells. Also, if you simply stop working a hot horse and put him straight into the stable he could cool down too quickly and get chilled. In fact it's a good plan to lead your horse home at the end of an active hack, especially if he's the type who may not calm down sufficiently while you are still on his back. Run up your stirrups, loosen the girth two holes (but no more in case the saddle slips and panics him), bring the reins over his head unless he is wearing a running martingale and then lead him home, walking on the left side of the road with you on his right and with your whip in your right hand for control of his quarters.

Getting fit

There is no mystique about getting a horse fit: it is a logical, fairly easy process which *does,* however, take time, and there are *no* short cuts. Because the whole body needs time to adapt to the increasing physical demands on it, any attempt to cut a week here or a few days there simply means that the time being given is insufficient; and increasing the work before the body has adapted will simply overstress it, and this almost always results in injury of some kind. A mild injury may even go unnoticed by a novice owner, but as it is aggravated by further work, it becomes more serious; and then the time you were trying to save has to be taken anyway for recovery, plus yet more time to make up for lost ground.

To get a horse or pony fit you gradually increase the length and stress of his exercise over a period of a few weeks whilst gradually increasing his feed at the same time. A sample programme follows, but it must only be regarded as a guide to give you an idea of how to go on with your own horse. If you have never worked through a fitness programme before, discuss it you your adviser or instructor; you'll both know your horse quite well plus his existing work and feed levels, and

During your hacks and fitness programme, be careful not to constantly nag your horse into going correctly and up to his bridle or he will soon become sour and resentful. Hacks should be enjoyable and used as much for relaxation and a change from schooling in the manège as for healthy exercise and fitness. Give the horse frequent periods of an easy walk on a long (not loose) rein unless you are in traffic or in any circumstance when the horse might play up, such as, for example, if he does not like farm animals.

will be able to adapt a programme. Also, having someone experienced and knowledgeable at hand is a great help when you are in doubt about how well your horse is responding to the programme, and need an on-the-spot assessment about when or whether to step up work and feed or whether to hang back a while.

Working an unfit horse too hard for his level of fitness can result in injuries for reasons similar to those described, and children's ponies as much as racehorses must be made fit for the work they have to do. They are all made of flesh and blood, and vulnerable to injury; although on the other hand, it is surprising just how hard a properly fit animal can work provided his work and exercise are sensibly balanced.

Most privately owned 'general purpose' horses today are in work for most of the year, and rarely get into a completely soft (unfit) condition as do specialist horses whose work is seasonal and who are given a long, out-of-season rest, such as hunters and polo ponies. However, you should know how to proceed with an unfit horse in case yours is ever laid off for any reason.

If you have both stabling and grazing facilities nearby and the horse is normally on the combined system, you won't have the hassle of bringing him up until he is completely off grass – this of course involves a complete and quite upsetting change of both diet (from grass to harder feed) and routine, not to mention the stress of depriving him of liberty. If you have no option, however, make a start in preparing him for the change by riding him in the field (no shoes needed) just at a walk for half an hour every other day, and giving him a small feed of no more than ½lb (.25kg) of hard feed such as coarse mix at the end of his exercise so his digestive system can gradually get used to it.

Do this for a week; for the second week ride him every day and for three-quarters of an hour to an hour, still walking but reminding him about changes of direction, going up to the bit, halting and standing and walking off smartly again, to keep him alert and listening to you.

After this you can bring him up from grass to the stables completely if you have to, without too much digestive upset. Get him shod, with lightweight shoes at first till he gets used to the unaccustomed weight on his feet again, and start walking on the roads. These early weeks of just walking really pay dividends later in a fitness programme, as they lay the foundation for faster work and give the horse's body a

Most horses enjoy loose schooling and popping over small obstacles and it makes a 'fun' change from being ridden all the time.

real chance to get tuned in gradually to physical effort. Give him a constant supply of hay or hayage and water, and feed the same concentrate feeds as for week two, but split into fewer, smaller ones perhaps bulked out with chop and soaked sugar beet pulp, sliced carrots and so on.

It's a good plan to start off a fitness programme just riding in the field. Shoes are not necessary for this and the softer ground makes fewer demands on the horse or pony's legs while his body is adapting to work again.

In week four start trotting, perhaps on grass at first, and just gently; in an exercise spell of 1½ hours now, give two five-minute gentle trotting spells. His concentrate feed may now have been gradually increased over this initial four weeks to about 3lb (1.5kg) of concentrates a day. You should be riding the horse on five days of the week, or more if you wish. If you cannot turn him out at all he should really be ridden every day, as it is bad management to keep a horse standing in his box with no exercise (euphemistically called 'a day off'), even if you do drastically reduce the concentrates on that day.

In week five you can keep the duration of the exercise the same but could increase the trotting to two ten-minute spells on the road, nothing faster than a working trot but making the horse go properly up to his bridle without unduly restricting him – make him work from the back end with his hind legs driving him forward, rather than allowing him to slop along on his forehand. If you feel he can safely take more concentrates, increase them by another ½lb (.25kg) a day, or more depending on the horse.

He should now be hardening up a little and slimming down (if he was fat) and working energetically and enthusiastically. In week six, put up the duration of exercise to 1¾ hours a day; introduce two steady canter spells of two or three minutes each, and lengthen the trotting spells a little.

85

By week seven, he should be on two hours exercise a day, preferably in two spells – although this may just not be practical for you, depending on your circumstances. His cantering should be at working speed (check with your instructor) with two five-minute spells, two 10-minute trotting spells and the rest walking, partly on a long rein to relax and as a reward, and partly going up to the bridle. He should now be what is described as half fit, and this is reasonable for average riding club work such as weekend shows or an instruction class (which, however, normally involves rather a lot of trotting); and now you could introduce a little jumping.

Once you have reached this stage – with concentrates on perhaps 5 or 6lb (2 or 3kg) daily, depending on the horse – you can take things further by increasing the severity of the work; you could increase the concentrates too, if required, but do take expert advice as a certain amount of skill and experience is needed to fine-tune fitness beyond this stage.

Overfeeding is dangerous, and a good rule is always to increase the work before the concentrates and to decrease the concentrates before the work. A combination of correct feeding, exercise and grooming to make sure the skin is clean and working well, can hardly fail to provide a fit, healthy horse physically. He will be much happier, however, if you can arrange some turning out for him. Horses on the combined system can simply have their hay ration reduced to compensate for the grass they will be getting. Yarded horses, of course, can be fed just like stabled ones. I believe mainly in giving hay to appetite, and allowing the horse to control his own intake.

No horse can be kept at peak fitness indefinitely, although many are kept at least half fit most of the year, and every horse appreciates a short holiday; this is achieved by a process known as 'roughing off' and 'letting down' a horse. It simply means a gradual increase in turn-out time, and just as gradual a decrease in

This Thoroughbred is in a suitable state of fitness and condition for active hacking and Riding Club work – not carrying too much weight yet not so lean and hard as would be required for hard work.

concentrates, exercise, grooming (body brushing) and, depending on the time of year, clothing worn.

If you are roughing off in spring or autumn, have a turn-out rug at the ready as these do lessen the shock of increased exposure to the weather; and in any case, you shouldn't start roughing off and letting down a stabled horse while the weather is worsening or still wintery.

Choose a mild spell and remove one blanket or under-rug, if worn, and start turning the fully stabled horse out for just about half an hour to an hour at the most, being extra careful if your grazing is at all lush and rich – though this is unusual for the run-of-the-mill horse paddocks which are often rather bare! If you are going to have to turn the horse out completely because your stables are too far from your grazing to keep bringing him in and out, start your roughing off process two weeks earlier by grazing him in hand, initially for half-an-hour a day until eventually you are spending your normal exercise time of two hours a day or so holding the horse in hand while he grazes. I know this sounds boring but it is the only real way to manage the change to grass feed without great risk to his digestive system. Try, too, to graze him for two one-hour spells a day rather than two hours all at once. He'll also need some exercise, of course.

If the horse has been quite fit, the whole process can take a good three weeks. At the same time gradually reduce the concentrates till he is having about 1lb (.5kg) of concentrates a day by the time he is going to be turned out. Reduce or cut out the body brushing altogether as he needs the natural oils in his coat as some protection against rain, and reduce his rugs to none by the time he is to go out; but if the weather is chilly *do* put on his turn-out rug for at least a week.

You have to use your common sense and keep an eye on the weather, and on whether or not he is feeling cold without the rug. If he is, his coat will stand away from his body, look dull, his loins and the base of his ears will feel cold and he could look pinched or even shivery. If he is cold and miserable he could lose condition considerably and may become ill. With proper facilities and a continuation of concentrate feeding in the field, if necessary, he will acclimatise eventually, but it takes time.

Lungeing is often used for short spells of exercise and can be valuable for allowing the horse to let off steam, within reason, and for encouraging him to go well in a nicely rounded outline like this. The horse wears protective boots, and the trainer a hard hat, gloves and substantial boots or shoes.

ABOUT FACE

If horses are allowed to travel loose in the back of a horsebox, it is normally found when the ramp is lowered to unload them that they have positioned themselves with their tails to the engine of their own free will. It does not take much imagination to realise why. It is not uncommon for horseboxes to have some rear-facing stalls, depending on the interior plan of the box, but in the United Kingdom at least, there are no rear-face trailers made at the time of writing. It is not possible just to change the interior arrangement of a trailer so that the horse can travel rear-face, because this will upset the whole balance of the trailer. The axles would need repositioning before this would be safe because the bulk of the horse's weight would then be at the back of the trailer instead of the front, which could cause it to tip backwards, raise the hitch and pull the back end of the car off the ground with it – not an amusing experience!

There are ways, however, in which you can ameliorate the dynamics of motion discussed: simply ensure that you or your driver drives with due consideration for your precious live cargo by accelerating and braking *very* gradually indeed so as to make changes of speed hardly noticeable, and by making all sideways movements very slowly and as wide as possible so as not to swing the horse around. It is also good to give yourself plenty of room between you and the vehicle in front, and to watch traffic lights well ahead so that you never have to brake suddenly. Avoid potholes in country lanes and tracks, and if driving over a field to the horsebox park at an event pick the most level route possible and *go slowly*.

TRANSPORTATION

Travelling in a horsebox or trailer is something that most of today's horses and ponies have to tolerate. It has gone out of fashion to hack miles to a show or meet, work all day and hack back. Shortish rides of two or three miles to an event are still common, but most people would try to get transport, if they had none of their own, for anything further than that.

There are two main ways of transporting horses: by trailer towed behind a car or Land Rover, or by horsebox. Some people are content to let their animals go in a cattle wagon with no proper partitions separating them at all, just a single bar or pole between them; this is quite inadequate (for cattle as well as horses, in my opinion) as there is nothing to stop them treading on each other as they try to keep their balance, and nothing to stop them being trampled on should they actually fall.

Horseboxes are by their very weight and construction much more stable in motion than trailers, but whatever is used it is the driving which matters: far better to travel a horse in a carefully driven trailer than a carelessly driven horsebox.

A very major reason for so many animals having a bad time when travelling is that they are obliged to travel facing the direction in which they are going. For several years now, scientific and practical evidence has been available to prove that horses travel least stressfully with their tails to the engine; this is partly because of the horse's physical build and natural balance, and partly due to his mental attitude.

Horses naturally carry about two-thirds of their weight on the forehand. When facing the direction of travel, the momentum created when the vehicle accelerates lurches the horse's weight back on to his hindquarters, which were never designed to take it; and this happens, of course, not merely when the vehicle sets off but whenever any adjustment of speed is made during the journey, which must be hundreds of times on a longish trip. The muscles of the back, quarters and hind legs brace hard to keep the horse in balance and upright and so are working constantly during a journey. Furthermore whenever the vehicle corners, changes from one lane to another, or negotiates a roundabout, not only is there usually a change of speed but the horse is lurched from side to side and back again, particularly at roundabouts. This means the horse must brace his hind legs sideways – and often his forelegs, too – in a stance which is completely unnatural; it also means he has to adopt it more or less constantly for the duration of the journey simply to stay on his feet.

The horse is therefore having to use his muscles in a quite unnatural and unaccustomed way for the entire trip, and some last for several hours. This not only leads to significant mental distress, but also overworks the muscles involved; this is tiring for the horse because of the energy needed, and means that by the time the horse reaches his destination and is required to work he could be already tired out – particularly if he has had an inconsiderate driver. Overworked muscles frequently cramp up, too; in mild cases this may be taken by the people accompanying the horse as 'a slight stiffness'. After a day's work the horse is again subjected to a journey to get him home again.

Another reason why so many horses are reluctant to travel is that the braking momentum lurches them forward on to the forehand; the horse counteracts this by again bracing his quarter muscles to get himself upright and avoid, in his mind, hitting his vital and sensitive head on the front of the box. All this may sound exaggerated, but if the driver is not extremely careful and considerate, it is not. To greatly lessen these effects drive as though you have no brakes.

Other reasons for poor travelling behaviour are lack of ventilation in the box which makes the horse feel under par, and the generally claustrophobic feeling of travelling in an enclosed space with often no way of seeing out, and no warning of what is going to happen.

Lead the horse *straight* down the ramp and stay level with his head or slightly behind it. It's safest to wear your hard hat and sensible boots or shoes, plus gloves, in case of mishap.

Getting ready

When preparing for a journey it is important to be sure that your vehicle – whether horsebox, or car and trailer – has recently been checked (and that it is regularly serviced, anyway) and is safe, that the tyres are all at the correct pressure, that the floor of the horse's travelling space is safe and well bedded down, ideally with six inches of damp sawdust to provide a really stable footing, and that you put everything you are going to need into the vehicle first. The horse goes in last.

For safety's sake the horse will need protective clothing, consisting of stable bandages over padding (or travelling boots), over-reach boots, knee and hock boots, a rug or light sheet to protect him against rubs, a tail bandage, perhaps covered by a tail guard which ties to the rug or roller, a headcollar and a poll guard.

Equipment needed during the day might comprise tack, work bandages or boots, spare rug, feed, bale/s of hay, containers, haynets, large container of water, first-aid kit, studs, plaiting-up kit, grooming kit, class entry form copies, directions how to get to venue, car park passes and admission passes, vaccination

Competing at shows can be great fun provided you don't make it the be-all-and-end-all of your life with your horse. Your relationship with him is much more important than how many rosettes and trophies you win. If you retain a happy, relaxed frame of mind your horse will pick this up from you and enjoy outings to different places.

certificates, your own clothes, and a shovel to skip out the box or trailer so the atmosphere, particularly in summer, does not become putrid inside.

If the day is hot when you arrive, park in the shade if at all possible, open all ventilation outlets if they are not already open, and do not leave the horse in the box if it is at all stuffy. It is far better to let him be tied to the box outside in the shade, as long as someone is always around to watch both him and your equipment (which can easily go walkabout at shows).

Do try to ensure that your horse receives his feeds as near to his normal times as possible, bearing in mind the times of his classes or phases. Few things, apart from a rough journey, make a horse feel off colour and bring on colic quicker than erratic and missed meals. Let him graze as well and keep a haynet with him whenever possible, and offer him water frequently. Remember also that many horses will do droppings in the box but will not pass urine (stale). Give him every opportunity to do so if you want him to be comfortable, rather than risk him staling with you on board at a highly inconvenient moment!

LOADING AND UNLOADING

These two manoeuvres cause more angst among horse owners than possibly anything else, particularly loading. The way to do it is simply to take the horse's lead rope without fuss, knowing absolutely that he *will* go in, (confidence, you see) and walk him in a straight line nonchalantly and confidently towards and up the ramp. Stay level with his neck and don't get in front of him, because if anything will stop him this will, particularly if you look back at him. If he does stop, stop with him, keep your back to his tail, talk to him but don't look at him, and say: 'Walk on'. Don't ever try to pull him forwards like a reluctant dog on a lead; it never works. If necessary take him back, walk two or three small circles near the bottom of the ramp and try again, when it probably *will* work.

BAD TRAVELLERS

Many horses and ponies are bad travellers, partly because they seem in general to be rather claustrophobic (some extremely so) and often because they have been frightened; they may have been thrown around in the back of a vehicle, or may even have had a fall or some other bad experience. Perhaps when they have shown their dislike and fear of having to travel, they have been beaten to get them inside the trailer or box – and this, of course, has simply made them worse, as they then associate vehicles with a good thrashing. No wonder they are dubbed 'bad travellers', but so often it is the fault of their handlers.

If it doesn't and the horse is starting to actually dig his toes in, a friend giving him a kindly pat on the tail may do the trick, or perhaps putting a forefoot on the bottom of the ramp. The horse may simply need a few minutes to weigh up the situation and see where he is expected to go, and will often then pluck up his courage and go in.

It helps to face the entrance uphill if possible, to lessen the slope of the ramp; and try to position the vehicle so that the sun or any other light is shining into the back and clearly showing the horse his route. If there are lights inside the box put them all on, and have the box and ramp well covered in bedding, haynets hanging up and so on, so it looks normal and even inviting. Tempting the horse in with titbits or buckets of food often helps, and so does leading another horse in first, particularly if it is his friend. One thing is for sure – if you get uptight and start panicking, rushing, getting annoyed and even shouting, you will make the horse much worse, not better. He will quickly pick up your vibes and think there's something to be afraid of, and naturally won't load.

If you are really having trouble and it is absolutely essential that the horse goes in, call in expert help. It is possible to get most horses in, simply by having a careful assistant standing just behind and to one side of him with a stiff bristled stable broom, ready to prick him carefully under the tail when he stops. Most horses go in after this uncomfortable but not painful reminder. You could also try this method: fasten a lunge rein to each side of the box entrance and cross them behind the horse's thighs assistants standing on each side, holding one line each and gradually tightening them behind him.

This horse has decided not to go in after all, maybe because it is rather dark in the box and he cannot see properly where he is going, maybe because his handler has gone too far in front of him or maybe because he is remembering a bad experience when travelling in the past. He may need a minute to collect his thoughts and gather his courage, or need taking down to try again.

Dr Sharon Cregier of Canada (a personal friend) has done much of the research into transport of horses, and I know she regards these last two ploys as 'abuse from behind'. Abuse from behind is a major reason for horses playing up when loading, but there is abuse and abuse. Done quickly and firmly with no more ado than strictly necessary, these two methods do certainly reduce the anxiety and time needed to get a sticky horse quickly and quietly into the box. It's also true that some horses who have had a bad time in the past continue to play up even when driven regularly by a really good driver, and are almost certainly doing so out of habit.

My version of abuse implies thrashing a horse and generally mistreating it, a procedure which does not, in my experience, actually succeed. It simply makes the horse more and more terrified and resistant, and confirms in his mind the fact that travelling and everything to do with boxes and trailers is truly horrific – and you cannot blame him.

Unloading rarely causes such problems as loading, but can still be slightly problematical. In some vehicles the horse has to actually back down the ramp because it has been travelling facing the front and has no room to turn round. Some trailers have a front ramp down which you can simply lead the horse, but the more poorly designed ones do not.

If you can lead the horse out face forwards, untie him while someone lets down the breast bar, hold the leadrope firmly and just go with him. Be prepared for a rush as many horses hop down quite quickly. Get him used to the word 'easy' to slow and calm him down, and use it. Lead him straight down and let him walk about once you have landed.

If backing out, allow the horse to turn his head slightly to one side so he can actually see where he is going. It might help to have someone put a hind hoof on the top of the ramp so the horse can judge how far down it is and what the slope is. Don't rush the horse; say 'back' encouragingly, and have someone on the ground – preferably at both sides – guiding him down with friendly hands on each side of his quarters (not his tail or he might stop). If he hesitates, a slight push on his breast might help. Stay calm and let him take his time.

If ever you have real trouble, do call in expert help; there's usually plenty of it around at horse shows, but try to be sure the person really *is* an expert. Whatever happens, never let anyone thrash your horse; I can assure you it doesn't work, and you will almost certainly turn the horse into a bad traveller for life.

STABLE HYGIENE

Stables (and horseboxes) contain a lot of organic matter, since horses constantly stale and do droppings, sweat, rub, lie down and generally live; they are therefore an ideal breeding ground for fungi, bacteria and viruses. After any illness and at least once a year anyway, they should be thoroughly cleaned out and disinfected and allowed to dry out completely before use again. This is in addition to the regular vacuuming or brushing down of rafters, and regular disinfecting of the floor.

There are various types of steam and pressure water cleaners on the market, and also firms advertising in the equestrian and farming press who will come and do the job reasonably for you – and a very professional job they do, too. I feel it well worth the saving of time and hassle to let them do it rather than doing it yourself, particularly if you are a busy working owner – but the choice is yours.

Horses seem to dung regularly in their feed and water containers, and the best way to deal with this is to clean them out and then sterilise them with a baby's chemical bottle sterilising solution. This works, is perfectly safe, and leaves no horrible odour to put the horse off his feed.

It is useful to have in your equipment a bland antiseptic liquid such as Savlon, which can be used – just a dash – in the final rinsing water when washing your

DISINFECTANTS

Disinfectants should be chosen carefully. It is best to avoid some of the best known, very strong ones, also caustic soda and bleaches, as not only are they unnecessarily strong and toxic, they leave a smell which horses find really objectionable. There are several firms making special equine disinfectants which horses do not seem to mind, and they should be available from, or through, any good saddler.

horse. This helps a little to keep flies away, although for this task you really do need a long-acting fly repellent (often called 'residual' repellents and also available from good tack stores) for this important job. Coopers and Absorbine are both makes of good ones in the UK.

Another slightly unpleasant but important job is washing out the sheath of geldings. The thick, greasy discharge called smegma which accumulates in the sheath can eventually block it, which makes it impossible for the horse to let down his penis to stale resulting in urine running over his belly and setting up soreness and infection; furthermore the sheath itself can become infected with bacterial and fungal infections which can be extremely unpleasant and uncomfortable for the horse. The job should be done at least once a month, and more with some horses; very few object if you are gentle.

Get a friend to hold the horse and give him Polo mints or whatever he likes, or tie him up with a haynet. Have a pair of thin rubber gloves (if you wish), your 'back end' sponge, some medicated soap or baby soap and two buckets of hand-hot water, plus an old towel and some almond oil or liquid paraffin (called mineral oil in the USA).

Soap the sponge well and gently insert it and your hand inside the sheath, sponging round and round and right up inside, frequently taking the sponge out, rinsing it and repeating the operation two or three times, removing as much smegma as you can. Tell the horse he is a good boy all the time.

Rinse the sponge under the cold tap to kill the lather, then wet it in the second bucket of warm water and really thoroughly squeeze the water up into the sheath to get out every bit of soapy lather (You can use a hosepipe if you have a warm water supply available.) Do this several times till you are sure all the soap is out. Dry the area outside and around the sheath with the towel, then take some oil in the palm of your hand and put your hand inside the sheath again, smearing the oil all round inside. This will add to the horse's comfort and soften any stubborn, hardened smegma, plus making it more difficult for new deposits to cling. And that's it.

Not all horses will allow you to bring down their penis to wash it, and when this is necessary for veterinary reasons they may need sedating. It is not strictly necessary for this hygienic operation, but take the chance of looking at the penis when the horse is staling to check its condition, and give it a quick sponge, if possible, to remove any accumulated grit, bits and lumps of smegma, which often accumulate on it. Sometimes smegma forms into a hard lump near the end of the penis and must be taken off – quite easy. When the horse comes to trust you he will permit attention to this very sensitive and private part of his anatomy without fuss, and will feel all the more comfortable for it.

One more important task: feed containers and utensils should be scrubbed out daily with clear water and an old dandy brush or scrubbing brush, particularly those used for coarse mix or damped feeds as these cling to the sides of containers and eventually harbour germs. Water vessels become slimy quite quickly if not scrubbed daily, and it is tempting with automatic waterers not to do this but simply to check the supply. Scrub them, rinse them thoroughly and refill.

Your first-aid cupboard must obviously be kept clean, and all tops kept on medicines, ointments and liniments, if used. Equipment such as scissors, lint and bandages must be kept in a clean box or tin within the cupboard to stop dust and bits getting into them.

General common sense will tell you what to do once you have a reasonable amount of experience, and it is far easier to keep something clean than to leave it for weeks or months and then face a Herculean task putting it back in order.

You can buy special granules to sprinkle on your stable floor before bedding down which are claimed to help absorb unpleasant gases given off by rotting organic matter (droppings, bedding and urine). Deodorising cat litter, bought in economy-size sacks from a pet shop, also does a good job. Whilst not a substitute for good management, you may find these useful if you have to keep your horse, say at livery or in rented accommodation, in a stable which you feel is not as well ventilated as it could be, to help minimise stable gases and odours.

The *raison d'être* of the leisure horse is that it should provide pleasure to all concerned. However, any horse whose health, temperament, and 'trainability' is constantly being compromised due to poor or inappropriate feeding and nutrition can be a constant headache rather than a pleasure to own and handle. This is not just as simple as inadequate or excessive feeding, because the safety of both horse and rider can become an issue if the horse is over-excitable, or constantly fatigued, or accident-prone due to unsound nutrition.

In addition to this, any horse who becomes ill or has an accident due to incorrect nutrition, or whose convalescence after an unrelated illness is delayed due to wrong feeding, will inevitably cost the owner considerable time and effort, quite apart from its own poor quality of life.

By taking the horse out of its natural 'environment' (and in this context keeping a horse in a field or paddock is far from natural) you take over responsibility for its care, nutrition and exercise, and it is in the interest of you both that owners and handlers should have some understanding of how to feed the individual horse for the work required of it.

Whether you simply want to plod around country lanes or plan to compete in an international endurance ride, the way you feed your horse, and what you feed it can have a fundamental effect on the degree of enjoyment you obtain, *and* the degree of enjoyment the horse obtains, out of the whole proceedings.

THE DIGESTIVE SYSTEM

So, the first stage is to get a picture in our minds of just what happens to the feed once the horse has taken it in. The first thing to remember is that when we feed the horse we are also feeding billions of micro-organisms that live in his gut, and these in turn produce substances, including vitamins, which can be utilised by the horse. If we consider these micro-organisms in the way we feed the horse they can be extremely beneficial to the horse's health. However, if you upset them by making sudden dietary changes, or by introducing other stress factors which will be discussed below, they can have a profound and at times disastrous effect on the horse's health, and this may be manifested as scouring (diarrhoea), colic, laminitis (founder), temperament problems particularly hyper-excitability, and so forth. They may even be implicated in such problems as 'filled legs' and cracked hooves, and even some forms of cancer.

Let us follow the journey of a mouthful of mixed feed as it passes down the digestive tract or gut.

Effectively, the gut or GI Tract (gastro intestinal tract) is a long tube whose entrance is at the mouth and exit at the anus. The food itself passes through this tube without actually entering the body, but during the course of its journey it is acted upon physically, by the churning and squeezing action of the muscular walls of various parts of the tract; chemically, by acids and enzymes which are secreted at various points in the tract; and microbially, *ie* fermentation by bacteria and protozoa in the hind gut. All of these processes help in the release of the various nutrients which may then be absorbed into the bloodstream for distribution to the organs and tissues, providing the nourishment required for body maintenance and work. Those parts of the originally ingested food which have not been digested and absorbed are excreted at the anus as droppings, so the bulk of the material in the droppings has never, in fact, entered the body; this is unlike the material excreted via the urine and sweat, which has been processed within the body in some way.

In order that the food can be broken down efficiently, it needs to be prepared in an appropriate fashion, which may include cooking in some way, grinding or crushing, or some other form of processing, before it reaches the horse; and it may

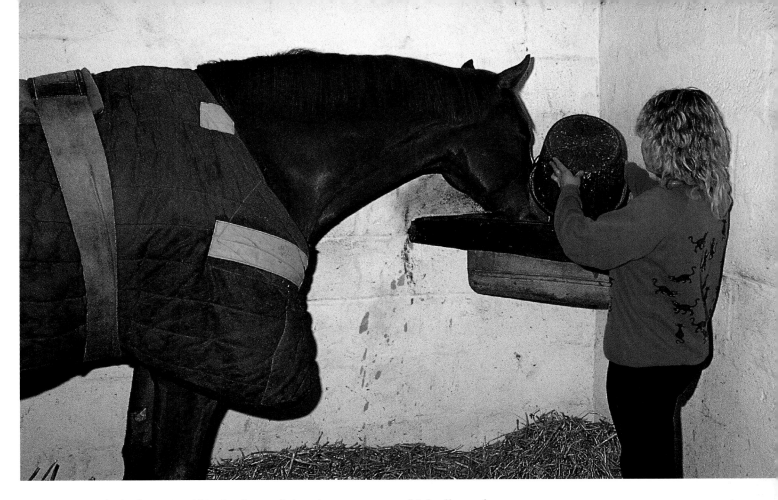

commence with the horse itself in the form of chewing, a process which allows the digestive juices to enter through the tough cell walls and commence breakdown of the foodstuff into its constituent nutrients. In order for this to be adequately performed it is vital that the horse's teeth are in good condition.

Horse's teeth continue to grow throughout life and if for any reason there is uneven wear, perhaps due to a slight misalignment of the upper and lower jaw, the chewing action may be impaired which will mean that the feed is inadequately ground in the mouth. Uneven wear of teeth can also lead to problems such as finding an appropriate bit, head shaking, sore mouth and even tooth abcesses, and must be dealt with as soon as it is noticed. To this end, teeth should be checked at least once a year, and every six months in younger and older horses.

Feed will also be wasted if parasites, including worms and bots, are consuming their unfair share, and perhaps more importantly, if parasite damage to the gut wall causes blockages and death of areas of tissue, preventing the uptake of nutrients from the feed. As many as 90 per cent of all colic cases have been related to worm damage, particularly that caused by migrating red worm larvae, and as many as 90 per cent of premature deaths in horses can be related to colic as a contributory cause or factor. Therefore, it is vital to worm all horses regularly and *frequently* with anthelmintics (wormers) – including stabled ones, broodmares and youngstock – using a programme compiled in consultation with your vet.

The importance of looking after the microbial population in the gut has already been mentioned; factors which disrupt their balance, along with gut motility (movement) and acidity, are frequently the fault of the horse manager, who should always be aware that disturbing this harmony can have a devastating effect on a horse's health.

These micro-organisms are important because they help break down the fibrous plant cell walls which 'protect' the nutrients within the plant cells and render them unavailable to the horse. One reason why we feed little and often is that

Always take particular note of whether or not your horse cleans up (eats all his feed) after each meal. Appetite is a good sign of his state of health. If he is normally a good eater and he starts leaving food, for instance, you can suspect trouble. However, some animals such as high performance horses and racehorses on high concentrate rations often go off their food and lose interest which indicates that a change or a rest is needed. They may also not clean up after a hard event or race when very tired.

THE DIGESTIVE PROCESS

To summarise, the digestive process is a highly complex chemical and physical process which is only beginning to be fully understood. The basic stages are as follows:

1 The food is selected by the horse, and some horses are better at this than others, which is an important aspect when considering grazing in areas where there are poisonous plants and so forth. It is taken in by the lips and, in the case of forage, bitten off and then chewed by the teeth. Saliva is released from ducts in the mouth and the digestive enzymes it contains commence action on the chewed food as it passes down the gullet. The whiskers are important sensory organs which are used to detect food, especially in the dark, and if possible these should not be cut or singed off, particularly when the horse is kept outdoors or has defective eyesight, as this deprives him of an important part of his food selection ability.

2 The bolus of food, which has been chewed and mixed with the saliva is pushed down the gullet or oesophagus by way of muscular contractions called peristalsis, to the stomach.

3 The stomach is a muscular sack which holds 8-15 litres of material (the equivalent to approximately one to two 2-gallon bucketfuls); its wall secretes a variety of gastric juices (water, mineral salts, mucous, hydrochloric acid and pepsinogen) which are mixed with the food by the churning action of the muscular walls. The food may stay in the stomach for as little as twenty minutes.

4 After the stomach the gastro intestinal tract or tube narrows dramatically to form the small intestine, which is 20-30 metres long (yes, really!), and made up of three regions, the duodenum, the jejenum and ileum; these may hold 40-50 litres of material. Digesta (ie the food being digested) are released gradually into the small intestine from the stomach, and are mixed with bile and pancreatic juices, which continue to break down the non-fibrous material. The proteins and glucose which have been released by this stage can be absorbed through the walls of the small intestine into the bloodstream, and their digestion is complete.

5 The rest of the digesta then passes into the large intestine, or hind gut; this comprises the caecum, colon and rectum, and is the home of the majority of the micro-organisms mentioned above. Further secretions of digestive juices also occur.

It is here that the action of the gut micro-organisms helps to break down the fibrous material, thus providing these micro-organisms with food for themselves; and they in turn produce volatile fatty acids (VFA's), which may be absorbed by the horse to be used as an energy source. As they die, these micro-organisms may also be digested by the horse, and the resulting nutrients absorbed and used to supply it with a small amount of protein along with a number of vitamins, particularly the B-group vitamins. However, it has now been realised that this cannot be relied upon as the sole source of B-group vitamins for young growing animals and high performance horses if they are to thrive and achieve optimum growth and performance.

This fermentation also tends to produce heat, and this may well be an important aspect of the horse's temperature control, particularly in cold weather when it may be especially important to feed adequate fibre in order to maintain an efficient fermentation process and thereby encourage internal warmth.

Unfortunately, the colon is somewhat badly designed in that it has a rather narrow right-angled bend or flexure which can become blocked, leading to 'impacted colic'. However, this problem can be completely avoided by supplying adequate water at all times, and feeding regular *small* feeds, and plenty of fibre; this is the type of regime the horse has evolved to thrive on, being basically a grazing 'trickle'-feeder.

6 Any remaining undigested feed is passed into the rectum and expelled via the anus, and the condition of the droppings can be a good indicator both of the type of feed being fed, and the health and well-being of the horse.

Some horses will occasionally eat their droppings; this is known as coprophagy and is an important practice for a foal who will eat the mare's droppings and thereby obtain a population of gut microbes. However in older animals or youngsters it may well be indicative of a number of problems, including a disturbance of the gut microbial balance, or possibly trace element problems, or gross protein deficiency.

Animals observed to follow this practice frequently and excessively should have their diets checked in detail, and if no obvious deficiency is found, a short course of probiotics may well put the situation right, although you should always consult an expert.

Wood chewing and eating soil can also be an indicator of nutritional disturbances including, in particular, shortage of fibre, sometimes minerals in the former case, and minerals and especially salt in the latter case.

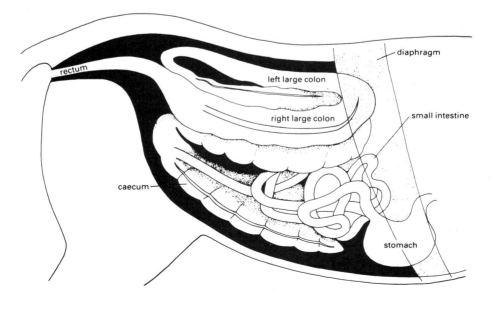

individually these 'bugs' may live for only a few hours, so to maximise their numbers and therefore ensure efficiency, they need to be fed on a regular basis. Another reason is that the horse has a very small stomach, and should not be overloaded with concentrate feed. A useful 'rule-of-thumb' is that you should give *no more* than 2kg (4½lb) of concentrates at one feed, and it is a good idea to add chaff to keep the ration 'open', unless plenty of other high fibre ingredients are included.

Different types of micro-organisms tend to prefer different 'substrates' or feeds, so one reason why we should feed a little of each feedstuff at each feed is to keep all the different microbial groups 'ticking over' and able to work at maximum efficiency; it is therefore not a good idea to give oats for breakfast, bran and sugar beet for lunch, and cubes for supper – give some of each ingredient at each feed.

If the ration is to be changed it can take at least three weeks for the microbial population to adjust, particularly from a high cereal or concentrate diet to a high forage ration (or vice versa); thus sudden changes in feeding can at worst cause massive disturbances, at best a loss in feed utilisation efficiency. Any such transition (including turn-out to grass or bringing a horse up again) should therefore *always* be done gradually over a period of 7-21 days, and if necessary gut stabilisers such as probiotics, sodium montmerrillonite clays, and *possibly* enzymes, may be used to help ease this transition, particularly in susceptible horses.

This aspect also applies to changing from one type or batch of hay or other feedstuff to another, from old crop to new crop cereals, or to a different brand of proprietary feed.

The nutritional quality, *ie* the nutrient content of the feedstuffs that are chosen will depend on the horse they are to be fed to, and ideally should be checked by laboratory analysis before a ration is formulated. However, even without a laboratory the horse owner can still check for visible and 'smellable' dust and moulds, as these can lead to allergic responses such as broken wind (heaves, Chronic Obstructive Pulmonary Disease – COPD) in both horses and humans; and the presence of fungal toxins may cause abortion in pregnant mares, infertility, loss of performance and very occasionally death, particularly in horses under stress. However, feedstuffs do not have to be *visibly* mouldy to contain moulds, which may be microscopic, so always buy the best quality feed you can afford.

COMPONENTS OF FEEDSTUFFS

Feeds are therefore normally ingested by the horse to provide a range and variety of specific chemicals known as nutrients, along with water; and also have various other non-nutritional functions, such as the physical action of roughage on the gut, and the digestive fermentation process as a source of body heat. As already discussed, at certain times the horse may ingest other materials, such as dung (coprophagy). Various drugs, too, may be provided orally by man and indeed, the horse may seek out its own medicinal plants (herbs, etc) in the wild.

The most obvious function of the 'normal' feedstuffs, however, must be to provide nutrients. Nutrients are biologically formed compounds which the horse acquires in varying quantities in order to function; the quantities required to sustain life are not necessarily the same as for optimum health or performance.

The nutrients themselves, apart from water, belong to a number of chemical groups: carbohydrates, fats and oils (lipids), proteins (and other nitrogenous compounds), minerals (including trace minerals) and vitamins.

Two other concepts are those of *energy*, which may be obtained primarily from carbohydrates and fats and *fibre* or *roughage*, which, whilst not being a nutrient, is a vital component of diet for its physical properties, and for providing a substrate for the gut micro-organisms to live on (see below).

Feeding is a highly specialised subject and research is throwing up new knowledge all the time. It's worth getting an equine nutritionist to assess your horse's codition and diet once a year as many less appropriately qualified people may not fully understand the current state of the art and may give outdated advice. Most reputable feed firms have a resident nutritionist who would help you free of charge if you use their feeds, or you could contact, in the UK, the Equine Nutrition Society, Woodlands, Cadsden, Princes Risborough, Bucks, HP17 0NB.

If you buy a new horse, do ask his previous owner exactly how he was fed – what ingredients, how much of each, what feed times – and try to purchase two or three weeks' supply of his existing food to gradually mix in with whatever you are going to feed, especially hay, to avoid digestive upset.

Do not underestimate the entertainment value of hay or some other roughage to your horse. Leisurely hours chewing his hay are the horse's equivalent to our reading, watching television, following a hobby or listening to music. Horses kept short of roughage frequently become restless and discontented. Outdoor horses, too, may well need hay if the grazing is poor at any time of year.

Energy

All living beings require a source of energy for 'maintenance', in other words to maintain body tone, to fuel basic life processes such as the beating of the heart and the activities of the lungs and the digestive system, and for the maintenance of normal body heat. Daily requirements of energy for maintenance in the horse are calculated based on its 'metabolic body weight'.

A source of energy will also have to be provided for muscular activity or work, the amount depending again on the horse's metabolic body weight, the nature of the work done and various environmental factors, along with the horse's temperament and the ability of the rider. Energy is also required for pregnancy, lactation and growth.

The primary source of energy in the horse from his diet is carbohydrate (see below) and to a lesser extent fats and oils, though higher fat diets are becoming increasingly popular as they are especially suited to certain equestrian activities. In addition, if there is excess protein in a diet, this can be broken down – though rather inefficiently – and used as a source of energy. This is an expensive way of providing energy, however, and it seems likely that excessive protein has a deleterious effect on performance – and the higher the level of performance undertaken, the more sensitive the horse is likely to be to this. Just as it doesn't make sense to feed the horse in such a way that it will be deficient in something, it *never* makes sense to feed it an excess of anything.

Fibre/roughage

Fibre or roughage in the diet is provided by the complex carbohydrates found in plant cell walls, namely lignin, cellulose, and hemicelluloses which may incorporate certain other compounds. The horse itself does not produce any enzymes which can break down cellulose, and lignin cannot be broken down either in the digestive process.

However, the population of gut micro-organisms in the large intestine is capable of breaking down cellulose – it breaks down the tough plant cell walls indigestible to the horse, laying open the cell contents (nutrients) for the horse's digestive processes. In return, the horse provides a nice, wet, warm environment for the 'bugs', and a regular supply of feed. The micro-organisms also produce vitamins, and are themselves a valuable source of protein when they die and are broken down and absorbed by the horse.

As growing plants become older, the cellulose cell walls become lignified, and progressivly less digestible even to 'bugs', which is why late cut hay is less digestible than early cut – although the level of nutrients may be as high, they will not be as available to the horse.

Inadequate fibre in the diet can lead to many digestive disturbances, ranging from blockages because the ration is not sufficiently open, through a whole panoply of disorders which relate to the disruption of the status quo of the gut micro-organisms. These can vary from filled legs to different types of colic, laminitis, temperament problems, eruptions and so forth.

THE NUTRIENTS

Water

Water is a most crucial nutrient, since more than 50 per cent of the horse's bodyweight (70-80 per cent in foals) is made up of water; less than a 20 per cent loss of body water will cause death rapidly. Water is required to establish and maintain the integrity or shape of every cell, bone included, and is a major constituent of every body fluid (blood, lymph). It also serves as a carrier for substances into and out of the body, as a constituent of saliva, urine and sweat, and is vital for the correct functioning of the metabolism and the digestive processes. Water plays a vital role in regulating body temperature, both by

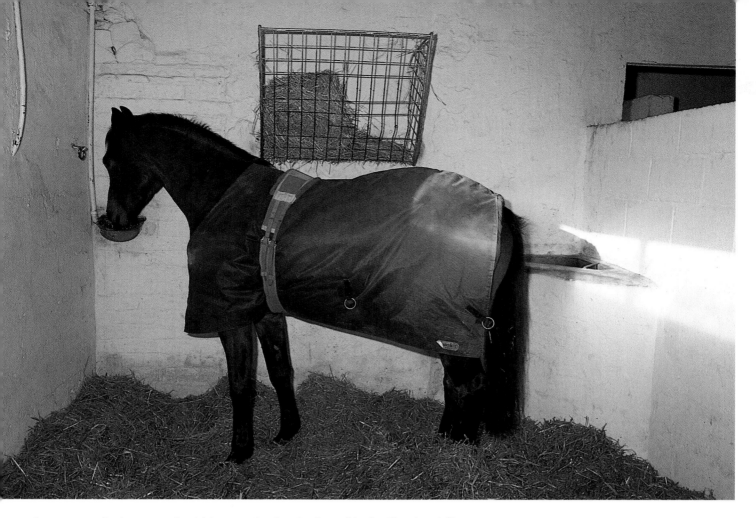

dissipating the heat produced in muscles by work and in facilitating adjustments required due to changes in the ambient temperature. The hotter this is the more the horse sweats and the more water he loses, although the extent to which this happens will also be governed by the ambient humidity, and the 'wind chill factor'.

If the horse sweats excessively, or has an inadequate supply of water, or both, he becomes dehydrated, and as this happens he can overheat, and the blood supply to the muscles may be reduced (ie the peripheral circulation), so that the waste products being produced during exercise are not removed, and oxygen and nutrients are not delivered efficiently, leading to fatigue or cramp-like symptoms. This is why it is a particularly bad practice to deprive the high performance horse of water for extended periods prior to and during competitions.

Water is also needed to lubricate the joints and eyes, transport sounds and carry nutrients around the body in blood and lymph which bathe the cells.

Broodmares will have an increased water requirement for milk production, and the milk is, at first, the only source of water for the foal.

Typically, a stabled horse requires 20-40 litres (5-10 gallons) of water per day. The amount it actually drinks will depend somewhat on the amount of moisture in its feed, and the horse at grass will be expected to drink less as it receives moisture from the herbage. However, dehydration is a frequent cause of colic in pastured horses in winter, especially when there is snow and/or water sources are frozen.

Change the water frequently rather than just topping it up, and scrub receptacles on a regular basis. Water takes in ammonia from the atmosphere (from urine) and becomes stale, 'flat' and, therefore, unpalatable, and in some cases even a thirsty horse may still not drink it. The use of electrolyte drinks is discussed below, but it should be noted that if you are giving one of these you should always offer plain water at the same time. This is because the horse has an

Automatic waterers are a boon in a busy yard with many horses but some have the disadvantage of your not being able to tell how much the horse is or is not drinking. As water intake, like food intake, is a sign of health or sickness, it's a good plan to use the type which has a meter so you can check on this. Also get the type with a recessed plug in the bottom which can be easily drained and cleaned.

Energy in itself is not a nutrient, but it is a vital component obtained by the 'burning' or 'oxidising' of 'nutrient fuels' in the feed. These fuels are largely carbohydrates, although to a lesser extent fats and oils, and excess protein in the diet may be used as an energy source.

Many horse owners tend to confuse the terms *protein* and *energy*, but these are by no means interchangeable. Protein *can* be used as an energy source if excess levels are fed, but energy can *never* be used as a protein, although it is needed for protein synthesis. Muscular activity is fuelled by energy and *not* by protein: energy is 'used-up' by muscles, whereas protein is not, so when work or performance levels increase, so do energy requirements. However, once the horse is mature it does *not* need large amounts of extra protein for extra work. Protein requirements for wear and tear are comparatively extremely small, although they may become slightly elevated in elderly animals, or those which are convalescent and less efficient at utilising the protein in their diet.

Energy is required for virtually all life processes and is effectively the most important component of the diet from a practical feeding point-of-view. Once you have got the energy level right it is relatively easy to adjust the rest of the diet to bring it into balance.

The level of energy in the diet is measured in most parts of the world in terms of Digestible Energy (DE), which is measured in Joules (which could be described as metric 'calories' – 1 Cal = 4.184 Joules). You can tell whether your horse is receiving adequate energy for maintenance of its health and condition by monitoring its bodyweight regularly. Assuming the horse is not working, as long as it is neither gaining nor losing weight and is of an acceptable body condition, then it is receiving adequate energy for maintenance.

inbuilt 'sense' for electrolytes, and if he doesn't need electrolytes but does need the water, he may go thirsty rather than drink the other electrolyte drink. The force feeding of salt or electrolytes, including the use of paste electrolytes, can in fact lead to dehydration unless the horse is very closely monitored, and it is almost invariably better to supply these in the form of a free-choice drink.

Carbohydrates

Carbohydrates are made up of carbon, hydrogen and oxygen, and in their simplest form occur as sugars particularly, for example, in fresh spring grass and molassed sugar beet pulp, whereas more complex materials such as starches are the main storage form of carbohydrate and occur for example in cereals (oats, barley, corn/maize etc). Fibrous feeds contain even more complex carbohydrates (cellulose and hemicellulose), which must be released from their relatively indigestible state by the action of the gut micro-organisms if the horse is to utilise them.

The very high and rapidly changing levels of soluble sugars in fresh spring grass can upset the digestive micro-organisms to such an extent that the horse or pony may scour, or even become endotoxic and suffer such metabolic disorders as laminitis (founder, or fever-of-the-feet). This can be partially avoided both by restricting grazing, and by supplying a good source of fibrous feed (hay or straw, or chaff) whilst the horse or pony is still at grass, to 'dilute' the effect of the sugars.

Fats and oils (lipids or lipins)

These are two and a quarter times as rich a source of energy when compared to carbohydrates, on a weight-for-weight basis (as every long-suffering dieter will know!) and they also provide a means of storing fat-soluble vitamins (A, D, E and K), and a supply of essential fatty acids. A certain amount of fat is present in most feedstuffs at relatively low levels but may be added as fish or vegetable oils, and the latter are particularly useful when devising high fat diets for performance horses, particularly endurance horses and those requiring a great deal of stamina.

Protein

Proteins are intimately involved in all the activities which sustain the life of all living cells. Each tissue, organ or organism has its own particular types of protein – for example, the protein in liver is different from that in milk, eggs or grass. Protein is used to build up new tissues in the growing animal and the pregnant mare, and for replacing worn-out tissues, particularly in elderly or severely debilitated horses.

Proteins are made up of amino acids, and there are 25 amino acids known in nature, 22 of which are required by the horse. These 22 are sub-divided into two types – 12 are referred to as *non-essential amino acids, ie* non-essential to the diet because they can be synthesised by the horse itself.

The other 10 amino acids are described as *essential amino acids,* as it is essential that they be present in the diet because they cannot be synthesised either at all or in adequate quantities by the horse. So for optimum health and performance, adequate levels of these ten *must* be supplied in the diet, along with sufficient protein constituents for the horse to manufacture the rest of its requirements for non-essential amino acids.

However, don't worry too much, as it is fairly difficult to underfeed seven of these essential amino acids to a mature horse on reasonable quality feedstuffs; the amino acid most likely to be deficient in equine diets is *lysine,* and the other two are *methionine* and *tryptophan,* although different feeding stuffs will tend to be deficient in different amino acids.

The total amount of protein supplied to the mature horse in light work may not be crucial provided it has adequate supplies of the essential amino acids, and it is only as the level of work increases that it becomes increasingly important for him to receive the optimum amount of protein as either a deficiency or an excess could damage performance in terms of work, and for broodmares and youngstock. The *quality* of protein is also important, and a diet containing 10 per cent crude protein with adequate levels of essential amino acids, may in fact be more effective than a 20 per cent crude protein diet with unbalanced amino acids; it is also likely to be somewhat cheaper.

The crude protein content of the feed appears on the bag label of commercial feeds, but is not, in fact, a particularly good measure of the usefulness of that protein to the horse, as the example above illustrates. Its digestibility is also important – in an indigestible form it is going to be of very little benefit to the horse; again, this becomes more crucial for high performance horses, broodmares and youngstock, than for the family horse in light work.

The desirable lysine content for the overall diet is in the region of 0.7 per cent for foals and weanlings, 0.6 per cent for yearlings, 0.5 per cent for 2-year-olds, and 0.25 to 0.4 per cent for mature horses depending on the level of performance; and elderly horses probably benefit from more, say 0.6 per cent plus a good supply of methionine for coat and hoof condition. Elderly horses do not *have* to look 'moth eaten' and scruffy, and they don't provided their very specific dietary requirements are taken into account.

Mature horses and ponies require 8.5-10 per cent crude protein in the *total* diet, that is the forage plus the concentrate. The term 'as fed' refers to the feed in the condition it is in, in store eg *un*soaked sugar beet, although this would be soaked before feeding; whereas analysis of feed on a 'Dry Matter' basis facilitates the comparison of apparently widely different feeds, for example fresh grass and hay, as they refer to the feed value of those feeds with all the water removed. This allows you to compare like with like.

There is a limit to the total potential Dry Matter intake of the horse, which is one reason why high performance horses in fast work cannot be fed an unconcentrated diet (eg fresh forage feed) in vast quantities as it is impossible for them to eat sufficient Dry Matter to receive the concentrated level of nutrients that they require.

Don't hesitate to ask for suppliers and manufacturers for what you want in the way of products and information. If you do not understand the information given on a proprietary brand of feed write to the manufacturers or ring them up if your merchant cannot adequately explain it and persevere until you understand the product sufficiently to use it wisely. If offered a product with no information at all don't buy it until you have been given adequate understandable information about its analysis.

PROTEIN REQUIREMENTS OF HORSES AND PONIES
(on an 'as fed' basis)

The following are some rule-of-thumb guide-lines to the crude protein requirements:

Broodmares in the last 90 days of gestation	Require 11% in total diet
Lactating mares first 3 months	Require 14% in total diet
Lactating mares 3 months to weaning	Require 12% in total diet
Foal creep feed	Require 18–20% in total diet
Foals (3 months of age)	Require 18–20% in total diet
Weanlings (6 months of age)	Require 16% in total diet
Yearlings (12 months of age)	Require 13.5% in total diet
Long yearlings (18 months of age)	Require 11% in total diet
2-year-olds	Require 10% in total diet
Mature horses: light work	Require 8.5% in total diet
moderate work	Require 8.5% in total diet
hard work	Require 8.5% in total diet
Elderly horses (16 years plus)	Require 12–14% in total diet depending on condition

Total diet refers to forage plus concentrates.

The typical mature horse doing medium work is not likely to require additional protein supplement (such as soya, linseed, fishmeal or skimmed milk), although additional lysine and sometimes methionine may be beneficial, especially as the work rate increases. Linseed is a particularly poor source of lysine amongst the high protein content feeds.

Excessive protein levels can be counter-productive, as they may increase sweating thereby encouraging dehydration, and cause 'breaking-out' after work.

High protein feeds tend to be expensive and it is foolish to overfeed them. Work out the energy levels required in the diet first of all, and then adjust the protein if necessary to achieve the optimum levels suggested above.

Vitamins

Vitamins are vital substances which although required in only minute amounts, are essential to the health and welfare of the horse. Broadly, there are two types of vitamins, those which are fat-soluble and those which are water-soluble. The fat-soluble vitamins, A, D, E and K, can be stored in the fat depots of the body and in general a grass-kept horse or pony on *good* pasture should store at least enough to see him through until Christmas. However, supplementation may be necessary after this time to achieve optimum health until the spring flush of grass appears. On the other hand, stabled horses are likely to benefit from some degree of supplementation throughout the year, particularly when they are in hard work.

The other, larger group of vitamins are the water-soluble ones, and these cannot be effectively stored by the horse which means that they are required on a daily basis. They consist of the B-group vitamins and Vitamin C, (ascorbic acid) and also Vitamin K, which is generally regarded as a fat-soluble vitamin but is also available synthetically in a water-soluble form. On a high fibre diet, many of the water-soluble vitamins are adequately synthesised in the gut by the gut micro-organisms, although if these micro-organisms are upset, for example by antibiotics or a sudden change in diet, or if the horse is not very good at absorbing and utilising these vitamins, additional levels may be required; the high performance horse, broodmares and youngstock are also likely to have increased requirements. This is partly because – particularly in relation to the B-group vitamins –as their carbohydrate intake increases they require more vitamins to utilise the carbohydrate, and also because the metabolism is 'firing' at a higher level and requires the diet to be more finely tuned to facilitate this.

However, there is little to be gained by feeding *excessive* levels as these cannot be stored and the horse simply excretes them, which is both wasteful and could even be damaging in high performance situations.

Generally speaking, vitamins are destroyed by heating and cooking, exposure to air and light, damp and mould. For this reason, vitamin supplements should be added to hot feeds *after* they have cooled, and containers should always be re-sealed and stored in a cool, dry place. Do not purchase vitamin/mineral supplements which have been displayed in a shop window or stored next to a radiator or hot light-source.

Minerals

Minerals form the major constituent of bones and teeth, and are also widely occurring in body tissues and fluids where they play a vital role in biochemical reactions, which are fundamental to the life process. The *total* mineral content of a feed roughly equates to the 'ash' content on proprietary feed bag labels. Unfortunately, however, unless this figure is very low or high, it does not really tell us anything very useful about the quality of the feed, because it does not tell us which minerals make up the ash or whether they are in an available and utilisable form.

The minerals can be sub-divided into the major or *macro-minerals*, which are required in relatively large amounts, and the trace or *micro-minerals*, which are

It's advisable to get a nutritionalist to do not only a soil analysis of your paddocks but also a herbage analysis so that you know just what is growing and can estimate the effect grazing is having on your horse's diet. This is especially important with breeding stock and performance animals. Grazing is part of the overall diet and must be included in your considerations.

TRACE MINERALS

The *trace minerals* usually considered are copper (Cu), cobalt (Co), iron (Fe), iodine (I), managanese (Mn), zinc (Zn), and selenium (Se), and there is increasing interest in the proper requirements for germanium (Ge) though these have by no means been investigated.

Trace minerals are likely to be in adequate supply for grazing horses on reasonably good pasture, if a trace mineralised salt lick is provided, although in some parts of the country specific soils may be deficient in specific minerals, and it is sometimes necessary to supplement these. Additional levels may be required for stabled horses in hard work.

required in minute traces but are, in fact, no less important than other nutrients. Other terms include 'trace elements', and 'trace nutrients', although the latter could also be applied to vitamins and is rather vague. Major minerals are calcium (Ca), phosphorus (P), sodium (Na), potassium (K), chlorine (Cl), and sulphur (S).

Provided your horse's diet contains adequate levels of forage, which is particularly rich in potassium, you should only have to consider adding calcium, phosphorus and sodium chloride (salt) separately. Cubed rations may contain adequate levels of these provided they are not diluted with other concentrates when fed, but proprietary vitamin/mineral supplements rarely do, for various good, practical reasons.

A deficiency of calcium or phosphorus, or an imbalance between the two, may manifest itself as brittle bones, sore shins (star fractures of the cannon bone), poor bone growth in youngstock and so forth. The *ratio* of calcium to phosphorus is important, apart from the actual amounts: suggested levels for mature horses are between 1.1-1.0 and 1.6-1.0 of calcium to phosphorus, never less than 1.1 and preferably no more than 2.5-1.00. Rations which contain high levels of cereal products, particularly wheat-bran, are especially likely to have an unbalanced calcium and phosphorus ratio, and they also contain a substance called phytin (phytate, phytic acid) whch locks up calcium from the rest of the diet. This is why it is always a good idea to feed a mixture of cereals and other dietary sources such as sugar beet pulp or alfalfa cubes which are rich in calcium, or if necessary to add additional calcium eg in the form of limestone. It is also a good idea to use chaff instead of bran as far as possible.

As far as sodium and chlorine (salt) are concerned, supplementation is an easy matter. A basic salt lick, trace mineralised as appropriate for the area in which your horse is grazing, is usually perfectly adequate for non-working horses or those in light/medium work, provided it is available in both feed and stable; however, lactating broodmares or horses in hard work might need to have additional supplies, especially when they are on home-mixed rations, and may need between 25-100 grammes (approximately 1-4 tablespoons) of salt per day, *in addition* to a salt lick or a loose-salt box which should be available as a top-up when the horse requires it. Sugar beet is also a good source of salt, and where this forms a significant proportion of the diet, slightly less salt needs to be added.

The rule-of-thumb for supplementation is that it should be 0.5-1.00 per cent of the total diet by weight, taking into account both the amount of salt naturally occurring in the feed ingredients and the degree of hard work, ambient temperature and humidity experienced. Salt-deficient horses may become dehydrated. Use iodised table salt and *not* 'dishwasher' salt.

Magnesium is not likely to be a problem in most situations, although it becomes more crucial in the high performance horse.

Vitamin/mineral supplements

When considering how to prepare a balanced diet, it is important not to fall into the trap of considering individual vitamins and minerals, and indeed amino acids, in isolation. There are complicated interactions between them, and between other nutrients, many of which are not yet clearly understood or even identified. Well-known examples are the interactions between Vitamins D and calcium and phosphorus, and Vitamin E and selenium.

Horse owners are under increasing pressure from proprietary supplement manufacturers to use products which are quite often inappropriate or unnecessary, though it must be said that a horse on a mixed diet, and particularly on a 'British standard diet' of oats, bran and hay *may* need supplementation, particularly as work levels increase.

It is not possible to state categorically which commercial product is the 'best one' without knowing exactly what it is to be fed in conjunction with, or what type of horse it is being fed to.

The following are some guidelines which I use when deciding whether or not to use a supplement and when selecting which brand to use:

a. animals on good pasture are less likely to have depleted their stores of fat-soluble vitamins than stabled animals, and a simple cod liver oil supplement from Christmas to the first flush of spring grass may be all that is necessary. *Do not* overfeed cod liver oil as it is a very concentrated source of nutrients.

b. Make yourself aware of any specific deficiencies in the soil in your area and choose a supplement accordingly. Both grazing and forage plants are variable feeds and tend to be deficient in different minerals, so different supplementation will be required. You should also take into account whether both forages and concentrates are home-grown, particularly if you are in a deficient region or even one where excesses may be a problem, or whether the feed is brought in from outside.

c. The grass-kept horse in light work will probably only require a trace mineralised salt lick as a supplement, although if the land is poor or very chalky (chalk locks up trace elements) and you are not giving concentrates, you might consider using a feed block (range block) to provide the concentrated nutrients which might be required. Adequate water and forage should always be freely available when such blocks are used.

d. The stabled horse, whose feed intake is being entirely controlled by you, is likely to need more attention to supplementation. You should consider which of his nutrient needs may be missing or low in the feed you are providing. For example, stored hay will have lost most of its Vitamin D after six months, and the stabled horse will not be manufacturing much of its own as it is not exposed very much to sunlight. Calcium may be deficient on grass hay and cereal diets, the amino acid lysine on most diets, salt on most diets, and phosphorus might be a problem, especially if alfalfa (lucerne) hay is being used.

You should also consider whether any of the nutrients are being affected by the treatment of the feed eg cooking, the addition of proprionic acid and other preservatives, rolling, cracking or crushing, all of which will reduce vitamin levels; and whether the type of diet may have reduced the horse's capacity to synthesise its own water-soluble vitamins.

When looking at proprietary products, don't be taken in by gimmicks and creative advertising, and don't assume that the product with the longest list of ingredients is the best. And do *not* be tempted to buy more than you can use by the 'use by' date, as although the trace elements will remain, vitamin levels will gradually deteriorate.

e. The requirements of elderly horses for supplementation are often neglected – this is why many older horses look moth-eaten and miserable. If an older horse appears to be grinding to a halt under your prevailing system of management, consider adjusting it, and take into account the problems of old age before writing him off for good.

Broodmares and youngstock will almost certainly require salt, calcium and phosphorus and, unless they are being fed a fully balanced concentrate at appropriate levels, they are also likely to benefit from a good broad-spectrum vitamin mineral supplement and a good quality amino acid source.

f. Bear in mind that if one scoop of supplement is good, two scoops are not necessarily twice as good, and may lead to a ration that is just as unbalanced as if you were feeding no supplement at all. Take care to follow the feeding instructions, and check them when you buy a new container, as the manufacturer may have slightly altered the product and the instructions may have changed.

DO NOT feed more than one broad-spectrum supplement at a time, and if possible feed by bodyweight. As supplements vary in their nutrient density, do not use a scoop from one brand for measuring out a different brand as you will not get a correct dose.

As your horse grows up, matures and grows old his work and nutritional needs will change. The supplement you have been using for some time may no longer be the best for him. His work and living conditions will also probably change throughout the year, so regularly reassess the supplement you're using, if any.

Be wary of feeding a supplement with compound cubes or coarse mixes (sweetfeeds) which have already been supplemented, particularly if these are not being diluted with other feedstuffs. However, in some cases judicious additional supplementation may well be beneficial, but be guided by an expert.

A final word of advice: do not pour hot water onto a feed particularly after supplementation, as this will destroy the vitamins.

Manipulation of vitamin levels in the diet occasionally appears to benefit some horses, eg megadoses of Vitamin E, methionine, folic acid or biotin, and possibly Vitamin C, but these should only be practised using expert professional guidance in strict circumstances. Biotin is not a cure-all *per se* for every hoof problem, nor Vitamin E or tryptophan for nervousness. They all have their role to play but should be used in a balanced fashion. Many owners of performance horses seem to be obsessed by feeding high iron levels, or Vitamin B levels, to raise the 'blood count'. However, it is NOT possible to raise the blood count above the normal level for a horse by giving extra levels of these nutrients; in fact, you are more likely to make the horse dull and lethargic. Similarly, any horse that responds to a vitamin 'booster injection' must – unless he is suffering from overt disease – be on a diet which is so poor that he is not being supplied with his basic requirements. Thus, anybody who is dealing with a high performance horse who claims to be able to boost his performance with injections must have a particularly poor feeding management system; these injections are in fact only likely to be appropriate in the case of a sick or debilitated animal, or occasionally for supplying fat-soluble vitamins to animals which are kept extensively on mountain or moorland and not brought in to be fed.

In all aspects of feeding and management of the horse, it is better to work *with* nature and not against it, or ignoring it. Work with your horse and he or she will be better able to work with and for you.

The range of feed supplements available is vast and bewildering. Some are 'broad spectrum' containing a wide range of different nutrients, and some are specialised containing perhaps only one. Never use any supplement without seeking advice from a vet or nutritionist. You can also check with the maker of a supplement about its likely usefulness to your horse: a reputable firm will not encourage you to use a product just to make a sale as it cannot afford to get a bad reputation from possible adverse effects through your horse having been given an unsuitable (for him) supplement.

FEEDSTUFFS

Assessing the nutritional value and quality of feedstuffs

The *only* way in which you can ascertain the feeding value or chemical analysis (nutrient content) of a feedstuff is to have it analysed in a laboratory, although some idea of whether or not the feedstuffs you are using are as good as you expect may be obtained from monitoring your horse's bodyweight, condition and performance. You can, however, make an informed guess about certain aspects of feed quality, and you can certainly judge by its physical appearance and smell whether it is clean and in a suitable condition to feed.

Feedstuffs are generally divided into forages and concentrates. The concentrates may be home-mixed using 'straights', or compounded by a manufacturer (nuts/cubes/pellets/coarse mixes – these are also known as sweetfeeds), or a home-mixed blend of these. These 'straights' may be divided into cereal concentrates and non-cereal concentrates and ideally, to achieve a balanced ration, a mixture of materials from the two groups is necessary.

Forage

The forage includes bulky feeds such as fresh grazed herbage, hay, feeding straw, silage, tower hayage, moist vacuum-packed forage (MVPF) – either fermented such as Horsehage or Propack or non-fermented such as Hygrass – and silage (picked grass!). Chaff or chop may also be considered part of the forage ration, as can treated straw cubes such as Viton, dried grass and dried alfalfa (lucerne) although the latter two are more usually considered with concentrates.

Hay is dried forage, usually grasses and legumes, cut at a relatively mature stage so that, although the yield may be high, the maturity of the plant means that much of the nutrient content is unavailable as it is trapped by the woody fibre. This is in comparison to grazed grass when it is fresh, silage and hayage which are cut at a much less mature stage, and also grass meal and cubes, and alfalfa meal and cubes (lucerne), which are also cut at a highly nutritious stage but differ from fresh herbage and silages in that they are then dried down to become as dry as, if not drier than, hay. This has the effect of considerably concentrating the nutrients.

Hay should be sweet-smelling, crisp to touch and a good colour, and there should be no patches of visible mould or dust. It is essential to check the centre of the bale for such signs.

Barn-dried hay is usually of superior feeding quality as its production is less dependent on weather conditions, and this should, of course, be taken into account when balancing the ration.

For performance horses a 'ley' hay (specially sown crop) of ryegrasses or timothy, or a good quality meadow hay, are the preferred products. Meadow hays may also include beneficial herbs but particular care should be taken to ensure that no potentially poisonous plants are present.

Make sure to use safe hay feeding equipment. Tie haynets so when empty, they will be above the level of the top of the horse's leg – it is quite astonishing how many valuable competitive horses are injured every year because they get a foot caught in a haynet or other feeding device. An ideal method for feeding hay to stabled horses is to use a 'hay corner' type of unit; this avoids the problem of hay seeds falling in their eyes which can occur with high hayracks, and wastage when feeding on the floor.

Ensure that horses using field racks cannot get a foot caught if the rack legs cross over as a part of their construction, or bang their heads on a circular feeder.

Moist vacuum-packed forages are of two kinds: the fermented type and the non-fermented type. The fermented products can vary in feed value somewhat, and the vacuum must be maintained in the bag to prevent spoilage. It is, therefore, essential that the bags are not punctured in any way or gnawed by rodents.

FEEDING AND NUTRITION
Feedstuffs
Assessing the nutritional value
and quality of feedstuffs

Two different types of hay. Meadow hay *(above)* and the coarser, more nutritious seed hay *(below)* often fed to hard working horses.

This type of cattle hayrack can form a hazard for horses who are not usually as quiet as cattle and can easily knock themselves. The author prefers long racks running down one wall rather than free-standing ones.

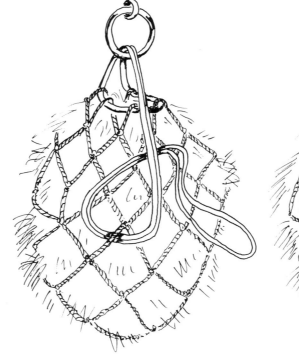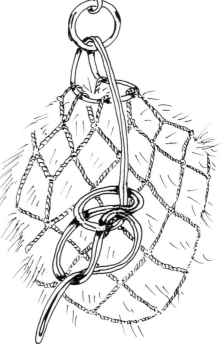

Pass the drawstring through the mesh about half way down the net and hoist it up again firmly.

Form an ordinary slipknot and pass the loose end through the loop of the knot, pulling the whole thing tight.

108

Although such products can be exceedingly useful for horses with respiratory problems, they tend to have a very high protein content, which applies especially to the lucerne/alfalfa types, but also includes many of the grass types. However, although their very high feeding value quickly satisfies the animal's nutritional requirements they are low in fibre and tend to be eaten quickly so many horses get bored and feel unsatisfied; it is therefore a good idea to feed another dust-free forage to 'dilute' the bagged forage, a useful example being a molassed chaff.

The other type of moist-packed forage, which may or may not be vacuum-packed but is usually packaged in some sort of plastic bag, is the non-fermented material which tends to be of a more consistent feed value and a lower protein content. It combines the benefits of 'reduced respiratory challenge' for moulds and dust which accrue from using a moist forage without the potential disadvantages of excessive protein levels.

All these products are of high feeding value and it is often possible to reduce concentrate feeds somewhat if they are used.

Silage is effectively pickled grass and has been made since Roman times. Properly made, it can be fed to horses under expert guidance.

Straw can be a useful feed for horses provided it is free of visible moulds and dust, although for the performance horse, it is most likely to be used to provide a source of additional roughage, perhaps as chaff, and a forage source when the horse is not actually working.

Oat straw from spring-grown oats is preferable as it has a higher feeding value, and in fact is quite often better than some so-called horse hays that appear on the market; barley straw may also be used, provided it is not full of awns. Wheat straw is not generally fed on its own although it may be used in compound cubes to increase the fibre level and 'dilute' the ingredients if they are too rich.

Dried grass, dried alfalfa/lucerne, molassed alfalfa chaff: these are available as meal, cubes, molassed chaff and, in some areas, wafers, and are highly nutritious; if used as a forage source they usually need diluting with chaff. Available in various grades, the meals are sometimes more palatable though dustier than cubes, but cubes are usually accepted if they are *lightly* soaked before feeding at first, a preparation which can be phased out once the horses are eating them.

When diluted these materials form a useful alternative forage for horses with respiratory problems, or when the usual forage sources are of poor quality or in short supply, but their particular advantage is that they are extremely useful when mixed with the concentrate ration in order to balance cereals, as they tend to balance out a number of vitamin and mineral deficiencies in cereals whilst providing protein of a reasonable quality and a good source of energy.

Concentrates

Concentrates are 'hard feed' and provide a more concentrated energy source (mainly starch) for horses in hard work, in cold weather, with impaired digestion (eg some elderly horses) or with a limited appetite (finicky feeders). For the high performance horse, even on good grazing or forage, some form of concentrate feeding will almost certainly be necessary because – as previously discussed – concentrate feeds contain less fibre than forage feeds, and when the horse's energy (nutrient) requirements increase, it simply cannot consume sufficient forage to obtain the energy potential it needs.

There are two types of concentrates: cereal concentrates and non-cereal concentrates. Cereal concentrates are the most commonly used concentrate feed for horses and comprise oats, barley, maize (corn), wheat products (micronised wheat, breadmeal, wheatbran, wheat feed), sorghum, various rice products and even millet. Non-cereal concentrates include sugar beet feed, dried grass or alfalfa

Haynets are a convenient method of feeding hay and minimising waste, but care should be taken in their use. They should be tied at horse's-head height and, in any case, so that the bottom of the net hangs no lower than the horse's elbow when empty to minimise the chances of his getting a hoof caught in the mesh.

A selection of feeds in common use. From top left clockwise: rolled oats, rolled barley, flaked maize and whole oats with, in the centre boiled linseed.

(discussed under forages above), and legumes and pulses such as soya, peas and beans; by-products which are mainly included in compound feeds including palm kernel meal, locust bean meal, cotton seed cake, ground nut cake, rape seed and canola meal, linseed cake, and soya bean meal, all of which are the meals and cakes left after the various plant oils have been extracted; plus other materials such as dried apple pomace which is quite widely used in some areas, along with linseed, vegetable oils, tallow, meat- and bone-meal and fish meal.

In general, healthy, mature horses on reasonable quality forage do not need the high protein levels to be found in soya, peas, beans, linseed, meat and bone meal or fish meal, although these may be desirable for breeding and youngstock. Many of these concentrates are processed in some way, usually in an attempt to make them more digestible, or in the case of soya or linseed, rendering them safe to feed. They may be rolled, flaked, micronised, extruded (gelatinised) or boiled, and in the future may also be irradiated.

Rolling to crack the seed coat and let in the digestive juices is necessary for barley but not for oats or maize which can be fed whole. Any benefits from rolling them may be lost as rolled cereals rapidly deteriorate in feed value.

Micronising is a way of cooking feeds rather like popcorn and they are then rolled before feeding.

Extruded (gelatinised) feeds are cooked to a porridge and flash dried.

Boiling cereals such as barley is generally a waste of time. There is little to enhance digestibility and, because boiling changes some of the constituents, if the boiled feedstuff is not given in every feed it in fact constitutes a sudden change in diet.

Cereal concentrates

Of these, *oats* are a favoured feed for horses. Grains should be large, plump, dry and clean and of even size in a sample; this will tend to mean that they have dried evenly and that there is less likelihood of spoiled and mouldy grain. Clipped oats have a lower fibre content because some of the husks have been removed, but are only economical if the digestible energy content is raised by more than the cost of the clipping. Oats can be fed whole, cracked, cut, lightly rolled, micronised or extruded. Whilst they are a good source of energy in the form of starch and have particularly digestible fibre as cereals go, they are very low in the amino acid lysine, and also in calcium having a poor calcium to phosphorus ratio, and they

ROLLED CEREALS

Remember that once cereals have been rolled or crushed they are dead, and dead things decay. If you think about it, if you plant a whole oat it will probably grow, because it is still alive. If you plant a rolled one it will simply rot as it is dead. Apart from this, vitamins are destroyed when air is allowed in by rolling or any other process, so don't roll or damage feedstuffs unless you have to, and only roll as much as you can use in two weeks, or less in warm weather. If you buy ready rolled cereals, you should check with your feed merchant that they are fresh.

You don't need a large feed room but it must be kept clean. Spillages in here (and in your horse's box or on the yard) attract rats very quickly so clear them up at once. Feed is best kept in galvanised metal bins.

contain phytate which will lock up calcium in the rest of the diet. The oil content is higher than in barley which means that oats spoil (go rancid) faster when rolled than does barley.

Barley can be fed cracked, rolled, cut, steam flaked, micronised, extruded or boiled but *not* whole. It is preferred by many horse owners to oats: it is much easier to buy a good sample of barley as more is grown than oats, and it tends to be less 'heating'. Barley has a slightly higher digestible energy content than oats so should not be substituted on a weight-for-weight basis, but should be incorporated into a balanced diet. The fibrous husk is much harder and less digestible than in oats which is why barley cannot be fed whole. Again, barley is low in lysine and has a poor calcium content, also a poor calcium to phosphorus ratio. Because of its lower digestible fibre content it is probably a good idea to break up the concentrate ration with chaff, dried alfalfa or sugar beet.

Wheat is not usually fed untreated as it tends to form a glutinous mass in the stomach which can cause colic. It is increasingly available both micronised or extruded, hence it can be fed safely, preferably with chaff. Higher in protein and energy than oats and barley, it should be introduced with care and the ration balanced appropriately. As with all cereals, it is low in calcium and the calcium to phosphorus ratio is poor.

There are several wheat products available: *wheatbran* is a source of very indigestible fibre and unfortunately has a *very* poor calcium: phosphorus ratio, also containing particularly high levels of phytin which can lock up calcium from

Different feeds. From top left, clockwise, coarse mix, molassed chop, crushed oats, rolled barley, pony cubes (in bowl), sugar beet cubes (loose) and bran. Sugar beet cubes are normally larger and coarser-looking than horse or pony cubes and grey in colour, pony cubes being usually green. The bran shown is rather fine unlike old-fashioned broad bran which is almost unobtainable now and was more nutritious because of former, less efficient milling processes. Bran is not an important horse feed today.

the diet, making it a very unbalanced feed source. Chaff (also called chop) is a much better balanced and more desirable roughage source. Because of bran's highly fibrous nature, bran mashes are highly indigestible and should not be necessary if the overall diet is balanced and in keeping with the level of work. And because they are so unbalanced, they constitute a sudden and therefore undesirable change in feeding, unless they are fed as part of each feed. For sick horses, or those laid off due to snow, increase the forage ration, and if you are already feeding dried grass, alfalfa and/or sugar beet, you can make a nutritious, palatable (revolting-looking) mash from these.

Regarding the normal working horse's diet on a rest day, if, for example, you are feeding three feeds per day, cut out the *weight* of feedstuff represented in one feed and divide the remainder to be given in three feeds as usual. You can also increase the forage allowance. For a Sunday rest day, ideally you would start this on the Saturday night's feed and go back to a normal feed on the Sunday night.

As a rule-of-thumb, if you *must* feed bran and are not giving significant amounts of calcium-rich feeds such as dried alfalfa cubes, alfalfa hay/chaff or sugar beet feed, you will need to add 50g of limestone per kg of bran for adult horses – more for youngstock.

Breadmeal is marketed in the UK under the proprietary name of Baileys No. 1 Meal, and Baileys No. 4 cubes (cubed breadmeal and molasses for palatability, with some added minerals), and is also popular – particularly in showing circles – in Australia and other parts of the world. It is an excellent source of 'non-heating' energy. It contains more protein than other cereals and has a higher lysine content. Again, the calcium: phosphorus ratio usually needs balancing, and as it is low in fibre it is usually a good idea to feed it with chaff, dried grass or alfalfa, NIS (nutritionally-improved straw) cubes or sugar beet feed.

It is safe to feed to children's ponies if they require additional digestible energy as it does not appear to 'hot them up', and is an excellent concentrated energy source for high performance horses.

Maize (corn) is sometimes fed in the USA on the cob, which increases the fibre level and is, therefore, safer; the grain may be fed whole, steam flaked, micronised, or extruded. Maize contains more digestible energy than either oats or barley and is lower in fibre, so it should be introduced very gradually and preferably be fed mixed with a fibre source such as chaff. Some varieties contain more lysine than other cereals, but again, the calcium and calcium: phosphorus levels are poor. It is a very good source of concentrated (starch) energy.

Non-cereal concentrates

Molassed sugar beet feed is often only used to damp feeds when, in fact, it can be used as a significant digestible energy source, as it is rich in salt and potassium, and an excellent source of calcium – quite apart from the physical advantages of damping down the ration. Up to 2.5kg per day (dry weight) may be fed, especially if there is a shortage of good forage, although the usual maximum level is 1.5kg per day (unsoaked weight). The shred or pulp should be soaked in *twice its weight* of cold water for 18 hours before feeding; the nuts must be soaked for 24 hours in *three times their weight* of cold water (one litre of water weighs 1kg). Do not use hot water to soak sugar beet pulp or nuts as the destruction of nutrients, and even dangerous fermentation, may result. The use of hot water *does not* speed up soaking time.

Of the *soya beans and soya bean meal,* full-fat soya beans cannot be fed to horses in the raw form as they contain a toxin. However, micronisation and extrusion processes have now made it possible to use full-fat soya – although it is unlikely that mature performance horses will benefit to any great extent from its use.

Concentrates don't automatically mean a tizzed-up horse. Investigate the 'non-heating', low-energy concentrates if your horse is susceptible to becoming too excitable on other concentrates. They are also ideal for animals prone to laminitis (if they have concentrates at all) and those confined to stables due to, say, illness, injury or frost and snow. Animals in light work may also do well on them.

OTHER FEEDS

Supplements such as *carrots* and *apples* are welcome treats to stabled horses. Carrots should be cut lengthways, and apples quartered. *Mangolds (mangels)* may also be used and are especially useful for horses with a tendency towards vices through boredom as they can be suspended from the roof for the horse to play with – though they must be where the horse *can* get them against a wall to bite if he wishes.

One material not yet mentioned is *hydroponic* cereal, such as so-called 'barley grass'; these are an extremely useful source of succulent feed and may be fed, roots and all, in the mats in which they are grown, or put through a chaff cutter and added to the concentrate feed. In the future I should like to see consideration given to adding micronutrients to the water in which hydroponic cereals are grown, to enhance their feeding value.

Soya bean meal is the by-product which is left after most of the oil has been extracted, so it is lower in energy content than full-fat soya (although relatively still high) but very high in crude protein content.

Linseed is the seed of the flax plant. It contains a fairly high level of rather poor quality protein but is rich in oil and mucilage. The flakes are again the by-product after much of the oil has been extracted. However, as linseed *must* be cooked before feeding, it is in fact far easier to obtain the protein from a better-balanced source and feed one to four tablespoons of *corn oil* or other vegetable oil per day to enhance coat condition (25-100ml).

Linseed must never be fed raw as it may contain hydrogen cyanide (prussic acid), a deadly poison. This can be destroyed by first soaking the linseed for 24 hours, then boiling vigorously for a couple of minutes and finally simmering it for a couple of hours (a great deal of time and effort for not much benefit!). You need around 2 litres (4 pints) of water for 225g (8oz) of linseed. Alternatively, it may be covered with water and cooked in a microwave oven for 20 minutes. One enterprising manufacturer is now marketing ready-cooked linseed ('Xlint') in the horse market, so if you are particularly keen to use linseed and can justify it economically, this is perhaps the simplest method.

After cooking, the resulting linseed jelly and tea should be fed and the tea can be further diluted to make it go further.

Salt, limestone, dicalcium phosphate and vitamin/mineral supplements

Try to divide these between each feed for the most effective use, but at least feed on a daily basis. Feeding any of these supplements once or twice a week is largely a waste of money and, although the horse will derive some small benefit, it would hardly constitute a balanced rationing programme.

COMPOUND FEEDS

These are rations which are manufactured by a feed compounder and may be in the form of a coarse mix (sweet feed) or cubes/nuts/pellets, which are similar to the coarse mix but the ingredients are ground and formed into pellets. A new development is 'formed' feeds made up of extruded ingredients.

With *complete diet cubes* the whole diet, forage and concentrates, is pelleted together, although in practice they are usually fed with a little chaff (molassed) for COPD-affected (broken-winded) horses, or long forage to prevent boredom. They may be useful for horses in transit, for competion horses being moved around a lot and especially for COPD-affected horses.

Concentrate cube/mixes are generally formulated to be fed safely along with hay, and diluting them with anything other than chaff (chop) or molassed chaff, which has been subtracted from the forage allowance, will unbalance them. Examples are those described as Horse and Pony, Hunter, Competition, Stud and Racehorse cubes – or coarse mixes (sweetfeeds).

Protein concentrates are compounds specially designed to be fed as a cereal balancer. They are a very useful idea and it is to be hoped that an energy balancer with less emphasis on the protein will become more widely used. This type of product is usually designed to balance the calcium, phosphorus, vitamins, minerals and possibly salt for the whole ration. Read the manufacturer's instructions for the brand you choose, however, as some may, in fact, suggest inclusion of a separate vitamin/mineral supplement.

Always make any transitions from home-mixed rations to compounds, or one brand or batch of compounds to another, gradually. Then wait until you have nearly run out before getting in a new lot.

If necessary, you can combine types of compounds such as Horse and Pony cubes with Racehorse cubes, but check with the manufacturers first.

ANALYSIS OF COMMONLY USED FEEDSTUFFS (As fed)

NOTE: These are average analyses and will vary from batch to batch of feed. These figures are offered as a guideline only.

	DE MJ/Kg	Crude Protein %	Crude Fibre %	Calcium %	Phosphorus %	Ca:P %	Lysine %
Hay, meadow	10.24	8.5	32.8	0.41	0.15	2.70:1	—
Good hay	11.0	10.0	30.0	0.45	0.20	2.25:1	—
Average Hay	8.50	3.0–8.0	38.0	0.2	0.30	0.66:1	—
Barley straw	7.07	3.7	48.8	0.24	0.05	4.80:1	—
Oat straw	8.17	3.4	39.4	0.25	0.07	3.60:1	—
Hygrass (Hayage)	11.58	13.1	39.2	1.00	0.30	3.30:1	0.16
Other Hayages vary considerably in CP, but tend to have a lower DCP value							
Dried grass meal/nuts 14% CP	10.66	14.0	26.0	0.90	0.30	3.00:1	0.80
Dried alfalfa 16% CP	9.70	16.0	24.0	1.40	0.20	7.00:1	0.64
Chaff (See Hay/straw)							
Sugar beet**	15.43	10.6	14.4	0.80	0.08	10.00:1	0.66
Oats	14.02	10.9	12.1	0.07	0.37	0.19:1	—
Barley	15.73	10.8	5.3	0.05	0.37	0.14:1	0.48
Maize	17.31	9.8	2.4	0.05	0.60	0.08:1	0.30
Breadmeal (Bailey's No 1)	14.90	15.0	1.3	0.24	0.28	0.86:1	0.30
Wheatbran	12.31	17.0	11.4	0.12	1.43	0.08:1	0.68
Non-heating H&P Cubes *	12.50	12.2	14.8	0.45	0.20	2.25:1	0.36
H & P nut from a national compounder *	10.0	10.5	16.0	1.4	0.65	2.15:1	0.25

© Gillian McCarthy 1990

*Compare these two and you will see how important it is to follow the individual manufacturer's instructions.

** *MUST* BE SOAKED BEFORE FEEDING (SHREDS: 12 hrs, PULP: 18 hrs, NUTS/PELLETS: 24 hrs. Do not soak longer especially in hot weather as fermentation may occur).

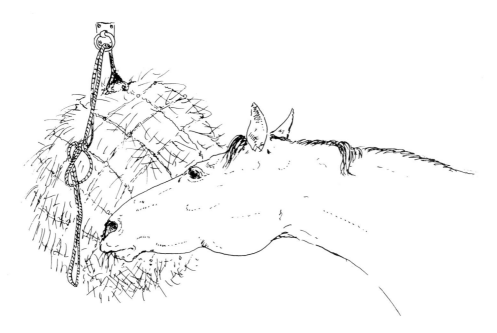

The horse can eat from the net at a safe comfortable height. If he pulls the loose end of the drawsting he can only tighten the knot and is in no danger of pulling the net down and entangling his feet. The knot on this drawing has been shown loose so that the arrangement can be clearly seen.

WORKING OUT A RATION

The following method has been adapted from the NRC method by Jeremy Houghton-Brown and is reproduced, with some amendments, with his kind permission.

Working out a feeding timetable such as this will probably not be the responsibility of the novice horsemaster; on the other hand it may well be relevant to him or her at a later stage, and may answer a few of the questions which a beginner would like to ask, but may not always dare!

Be prepared to sit down with a simple calculator for half-an-hour; time spent doing this will save time and money later!

Rationing in nine steps:

Step one: 'How big is the horse or pony?'
There are various ways to estimate a horse's bodyweight:

Method A: Weigh Tape eg Equitape. You read an estimation of weight directly off the tape.
Method B: Weighbridge; remember to deduct weight of tack and handler.
Method C: Tape Measure and Table
To estimate your horse or pony's bodyweight, take the girth measurement at a quiet time, right around the barrel; the tape should lie in the girth groove and just behind the withers. Read off the measurement when the animal finishes breathing out. The following tables are offered as guidelines, though it is best to check cob types against the tables on a public weighbridge to find which table applies to them. This table illustrates that just ½in (1.23cm) change in girth can mean 14–15lb (5–6kg) change in bodyweight, which may not be visible to the naked eye.

Once your horse or pony is fit, you should feed to maintain bodyweight throughout the season.

Table 1 Ponies

Girth in inches	40	42.5	45	47.5	50	52.5	55	57.5
Girth in cm	101	108	114	120	127	133	140	146
Bodyweight in lb	100	172	235	296	368	430	502	562
Bodyweight in kg	45	77	104	132	164	192	234	252

Table 2 Horses

Girth in inches	55	57.5	60	62.5	65	67.5	70	72.5	75	77.5	80	82.5
Girth in cm	140	146	152	159	165	171	178	184	190	199	203	206
Bodyweight in lb	538	613	688	776	851	926	1014	1090	1165	1278	1328	1369
Bodyweight in kg	240	274	307	346	380	414	453	486	520	570	593	611

(Tables based on work of Glushanok, Rochlitz & Skay, 1981)
Measure in kg: (50kg = 112lb)

Step two: 'How much hay a day?'
As rule-of-thumb, the following ratios of forage to concentrates, by weight, usually apply:

Mature horses in work

Light work (to medium)	25% concentrates to 75% forage
Medium work	50% concentrates to 50% forage
Hard work	60% concentrates to 40% forage
Hard, fast work	75% concentrates to 25% forage

25 per cent forage by weight is the absolute minimum for healthy gut function. In cold weather increase concentrate levels. If forage is of low nutritional value, concentrate levels will also need to be increased to balance the ration.

Broodmares

	Concentrates	Forage
1st 3 months after foaling	55%	45%
3rd month to weaning	40%	60%
Weaning to 90 days before foaling	25%	75%
Last 90 days of pregnancy	Gradual rise from 25% to 55% after foaling	Gradual decrease from 75% to 45% after foaling

Youngstock

	Youngstock destined to race at 2/3 yrs		All other youngstock	
	Concentrates/forage		Concentrates/forage	
CREEP FEED	100% plus milk		100% plus milk	
3-month foal	80%	20%	75%	25%
6-month weanling	70%	30%	65%	35%
12-month weanling	55%	45%	45%	55%
18-month long yearling	40%	60%	30%	70%
2-year-old	40%	60%	30%	70%

Step three: Work out energy requirement for maintenance
$$= 18mj + \frac{\text{Bodyweight in kg}}{10kg}$$

So, for our 500kg horse $= 18 + \frac{500}{10} = 68mj$ day

Step four: Work out energy requirement for production
For work per day, for each 50kg of bodyweight and mj of Digestible Energy (DE)

Light work	+ 1 eg One hour walking + 2 eg Walking and trotting
Medium work	+ 3 eg Some cantering + 4 eg Schooling, dressage and jumping
Hard work	+ 5 eg Hunting 1 day/week + 6 eg Hunting 2 days/week
Fast work	+ 7 eg 3-day eventing + 8 eg Racing

For lactation per day, for each 50kg bodyweight add:
For first 3 months   + 4½ mj of DE
For next 3 months   + 3½ mj of DE
NB All diet changes must be gradual, particularly for lactating and pregnant broodmares.

For pregnancy per day, add:
+ 12% for the final 1/3 of gestation or last 3 months.

For growth per day, add:
Youngstock over 1 year – feed at maintenance ration for their expected weight at maturity.
Up to 1 year provide 13 mj of DE per kg of feed, and feed to capacity.

We have worked out that the total daily energy requirement for one particular horse is 128 mj DE/day.
8.75kg of our 9 mj hay will supply:
9 x 8.75 = 78.75 mj DE (say 79) from hay
So our concentrate needs to supply 128–78.75 mj DE
128–75 mj DE = 49 from concentrates

We have decided to give 1kg/day (unsoaked weight) of sugar beet, which has a DE content of 13 mj (as fed but before soaking). So after 1kg of sugar beet, our concentrates need to supply:
49 − 13 = 36 mj of DE
Say we want to use oats and our sample has a DE of 12.5 (as fed)
36 ÷ 12.5 = 2.88 kg of oats

Step five: Adjusting the ration
We now have a ration of:
8.75 kg of hay rounded to 9 kg
1.00 kg of sugar beet (unsoaked weight)
2.88 kg of oats rounded to 2.75 kg

In practice, I would round this up or down to give manageable amounts, then find the percentages these represent. On a calculator, key in the following:

9kg Hay   = 9 ÷ 1 2 . 7 5

1 kg sugar beet = 1 ÷ 1 2 . 7 5

2.75 kg Oats   = 2 . 7 5 ÷ 1 2 . 7 5

Answer:
(total feed) % = 70.6% of total ration

(total feed) % =   7.8% of total ration

(total feed) % = 21.6% of total ration
             Total 100.0%

Supplying 128 mj DE/day

If your calculator has a memory, you only need to key in 12.75 once, then press:

| wt of hay | (eg 9) | + | memory | % | = |

On many calculators you will not need to press the = key.

Step six: Check the protein level
Find the protein level of your feedstuffs by laboratory analysis, or from tables, then:

Crude protein % of feedstuff x wt of feedstuff =
protein contribution of that feedstuff to the ration

For example, say our hay is 8% crude protein, sugar beet 11% and oats 10.9%. On the calculator.

Hay   8 x 7 0 . 6 % = 5.6 CP from hay
                                      +
Sugar Beet   1 1 x 7 . 8 % = 0.86 CP from sugar beet
                                      +
Oats   1 0 . 9 x 2 1 . 6 % = 2.35 CP from oats
                          Total CP in diet   8.8%

which is reasonably near our target of 8.5%

Step seven
In the same way, you should at least check the calcium, phosphorus and lysine levels, and as many other nutrients as you have time for, or are concerned about. I cheat a little, and use a computer to assess all the known nutrient requirements of each individual horse (including all the vitamins, minerals and several amino acids) but by hand that would take an inordinately long time. Of course, if you don't want to get bogged down, you could always feed hay and cubes, as guidelines for their ration levels will have been worked out for you by the feed compounder.

Step eight
Don't forget salt, and consider the vitamin/mineral supplement.
A sample table of nutrient levels in various feedstuffs appears on p115.
It is preferable to have your main inputs ie forage and, say, oats analysed for their actual feed values as they can vary enormously. I've asked 'experts' to pick out samples of 'good horse hay' and the analysis has ranged from 3 to 15 per cent crude protein! Oats can vary by nearly as much. People who won't buy cubes because they 'can't see what's in them' mystify me. You can't see what's *in* oats, ie you cannot judge their feed analysis just by looking at them. You can just see if they are a clean sample!

Step nine
Monitor the horse's bodyweight and performance and amend feeding accordingly. If he's overweight, work out the ration requirements for 5–10kg (no more) less than actual weight. When he's lost that amount, re-calculate the ration. Excessively rapid weight loss is not good for the horse, and 'starvation' slimming can precipitate hyperlipaemia, chronic laminitis, protein, vitamin and mineral deficiencies (requirements for these for maintenance remain virtually the same so when you cut down digestible energy for weight loss, make sure you aren't underfeeding the other nutrients), and also hormonal disturbances eg causing a tendency to put on weight even more easily (human slimmers should also be encouraged to avoid 'crash' diets for this reason), and again possibly chronic laminitis.
Keep weight loss steady and slow, and don't waste a winter slimming campaign by letting the horse put it all back on if he's turned out in the summer!

5 TACK AND CLOTHING

Kitting out your horse or pony with tack and clothing can easily cost as much as the animal itself and is a major expense frequently under-estimated by new owners – a decent saddle these days can cost about £300 and more. However, economies *can* be made, in quantity if not quality, by cutting down on unnecessary items.

It is possible, of course, to own your horse and virtually nothing else. There is nothing new in bareback riding and a well-mannered horse can be controlled with just a rope around his nose, but in our conditions today a saddle and bridle not only make for extra comfort, they can be regarded as essential to provide security of seat and control over the horse in traffic and among other animals.

If you have a horse or pony with a good deal of cob or native pony blood in him, or other hardy blood such as Cleveland Bay or Irish Draught, you may never need clothing if you are not going to clip him in winter. The same goes for bandages (apart from first-aid use), boots, martingales and all the other bewildering paraphernalia set out so temptingly in saddlers' shops and tack stores.

Let's take it, though, that you are thinking of buying your horse a basic wardrobe of saddle (with girth, stirrups and stirrup leathers), bridle (with bit and reins which are sold separately) and a good general purpose stable rug. Then we'll look at other items which could be very useful such as stable bandages, exercise and tail bandages, the most useful boots to consider, also turn-out (waterproof) rugs, and exercise rugs or sheets which can be useful for clipped horses in bitter weather.

SADDLES

Probably the best sort of saddle for general riding (flat-work, jumping and hacking about) is one with just that name – a general purpose saddle. There are also specialised saddles for dressage, showing, show jumping and endurance riding, and also racing. Side-saddle riding is also now very popular not only for ladies and girls but for anyone, male or female, who for physical reasons (such as arthritic hips) cannot ride astride.

Saddle-making is an ancient craft, and as equestrianism has spread in popularity, so the demand for special saddles and better design has come on in leaps and bounds to cater for all its different branches: the 'performance' riding sports such as eventing, dressage, show jumping, showing and long distance riding, and in some countries, the steady popularity of hunting with its associated branches of hunter trials, drag hunting and team chasing.

Saddles are traditionally made of leather, with various webbing attachments for girth tabs (straps), and horsehair or flock for padding the panel beneath the seat, all mounted on a beechwood frame called the 'tree'; synthetic materials are increasingly being used, however, both for the tree and for the other parts too. There are 'rigid tree' and 'spring tree' saddles. In the latter two lengths of springy metal are fixed from front to back of the saddle on either side so the rider has a more comfortable ride. These are more expensive than rigid tree saddles, but most people feel the extra cost is well worth it.

WHERE TO BUY

If you browse round saddler's stands at shows and other events, shops and tack supermarkets you may notice that some vendors appear to be knowledgeable and helpful while others don't know much about their merchandise and only want to sell it. Certainly your best plan is to buy from a retailer who is a member of the Society of Master Saddlers; the society has a distinctive crest which will be displayed on the premises and probably on the firm's stationery. Such retailers will have at least one qualified saddler on the staff who can advise you on the fit, quality and suitability of any item you are considering, and there should also be at least one 'horsey' person available for queries. This may sound obvious but, particularly in some of the tack 'superstores' which have sprung up, it's amazing how many of the staff don't know anything about horses but are simply there to sell the goods, which is no good to you at all.

Buying from a good firm not only ensures good quality merchandise and advice, but also a good after-sales service should anything go wrong, or if your tack needs mending or a periodical 'service', which is always a good idea.

These firms will often have a good stock of used tack, too. Secondhand tack of good quality is far better than poorer quality new tack; good tack, well looked after, lasts literally generations – poor tack breaks (and may cause accidents), goes out of shape and wears comparatively quickly. It is never a good buy. If you have the help of an experienced adviser or consultant, you may be able to buy good items very cheaply at auctions or from private sellers (though even some of these, remember, can be quite unscruplulous!) – but do be careful to take advice if buying from such sources as there is a lot of rubbish around, not only in terms of quality but dangerous and damaged equipment, too.

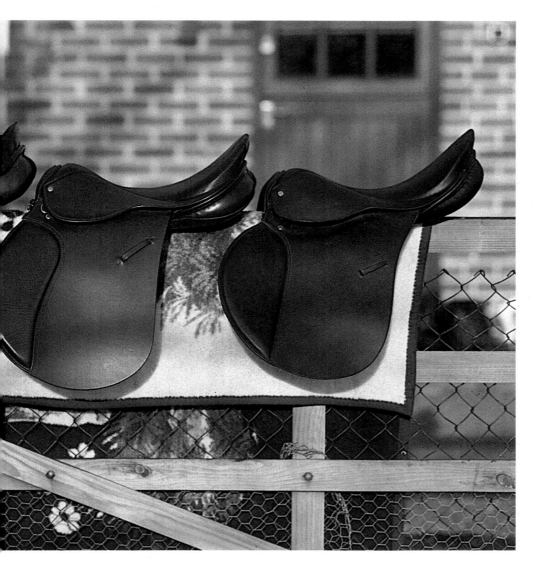

A selection of saddles, from left a British-style dressage saddle with long girth straps, a continental-style dressage saddle with short girth straps, a general purpose saddle and a jumping saddle.

UNSADDLING

To remove the saddle, hold it at pommel and cantle and *lift* it up off the horse's back. As you bring it towards you catch hold of the girth as it comes over the back and lay it with the inside (sweaty side!) over the seat. Sweat is easier to remove than mud, which can also scratch your saddle. Pull it off carefully and place it with the gullet over your left forearm, the pommel in the crook of your elbow.

PUTTING ON A SADDLE WITH A NUMNAH

(*left*) Pull the numnah well up into the saddle gullet and place it gently on the horse's back, well forward. Have the stirrups run up their leathers. The girth may be attached to the offside girth tabs and laid over the seat, or be fitted later. Slide the saddle back into position. If you get the saddle too far back, do not slide it forward again as this will ruffle the hair underneath and be uncomfortable for the horse. *Lift* it up and start again. (*right*) The saddle correctly placed with the pommel just over the withers and the numnah pulled well up into the saddle gullet for its full length.

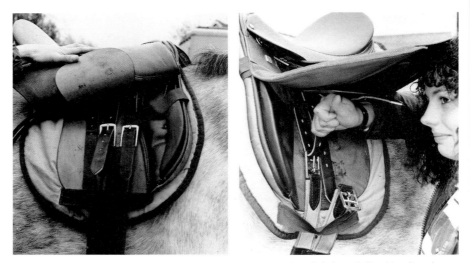

(*left*) The girth is fastened to the offside girth tabs, usually the first two, as here. If the horse is fit and lean it can be fastened to the last two, which will tend to help keep the saddle forward a little. The first girth tab is slotted through the loop on the front of the numnah to help keep the numnah forward, and the girth itself goes down through the loop near the bottom edge, again to help keep it in place. (*right*) From the near side, reach under and bring the girth up, making sure it is not twisted, through the bottom loop on the numnah, and fasten the first tab just tight enough for the horse to feel it. It is correct to rest the saddle flap on your head while doing this.

(*left*) The saddle girthed up on the near side, with the first tab through the numnah loop. This girth has elastic inserts (behind the bottom loop) for extra 'give' and comfort for the horse. (*right*) Correctly saddled up with a well-fitting saddle. The numnah is well up in the gullet and there is a clear space of air down the horse's spine. The stirrups are run up their leathers which they should always be when the rider is dismounted. This girth is an Atherstone, tapering behind the elbows for extra freedom of movement and comfort.

The style of riding and length of stirrup adopted dictates the style of the saddle – basically, the shorter the stirrup the more 'forward cut' the flap, the part on which your leg rests, because as you shorten your stirrup your leg bends and your knee comes further forward. For example in dressage, a long-ish stirrup is used so the saddle flaps are not forward cut; for general riding and jumping to quite a reasonable standard including cross-country jumping and in the lower levels of show jumping, the flap is moderately forward cut and the saddle is called a general purpose saddle; and in show jumping shorter stirrups are sometimes used, necessitating a more forward-cut flap, although the fashion now is for slightly longer leathers for greater control and stability. There used to be special saddles for showing which were – and sometimes still are – very straight cut to show off the horse's front. However, many people find these most uncomfortable and the trend now is more and more to use dressage saddles, although some saddlers do produce a specially made, comfortable, staight-cut show saddle.

Good, modern saddles have a 'central seat' – that is, they automatically position the rider in the centre of the saddle seat. The seat itself is deeper than in old-fashioned saddles, which makes for much more comfort and security. The centre should be the deepest part of the seat, too, for proper balance, otherwise the rider could be constantly fighting a losing battle to stay balanced and central over the horse's natural centre of gravity, which for all practical purposes is situated just behind his withers three quarters of the way down inside his chest towards his breastbone.

A modern saddle of whatever design will probably have knee rolls (of padded leather) on the panel under the flap for extra support for your leg position and knee when jumping, in particular.

TACK AND CLOTHING

Saddles
Putting on a saddle
Parts of a saddle

Parts of a saddle:
A   Cantle or back arch
B   Pommel or front arch
C   Gullet
D   Panel
E   Ends of girth tabs between panel and flap
F   Waist
G   Seat
H   Skirt, covering stirrup bar
I   Flap
J   Stirrup leather loop
K   Buckle guard to protect underside of flap from friction from girth buckles
L   Sweat flap, originally to protect flap from sweat in old-style, short-panel saddles but superfluous in modern saddles.
M   Girth tabs for attaching girth
N   Point of tree (inside saddle)
O   Knee roll for support and stability

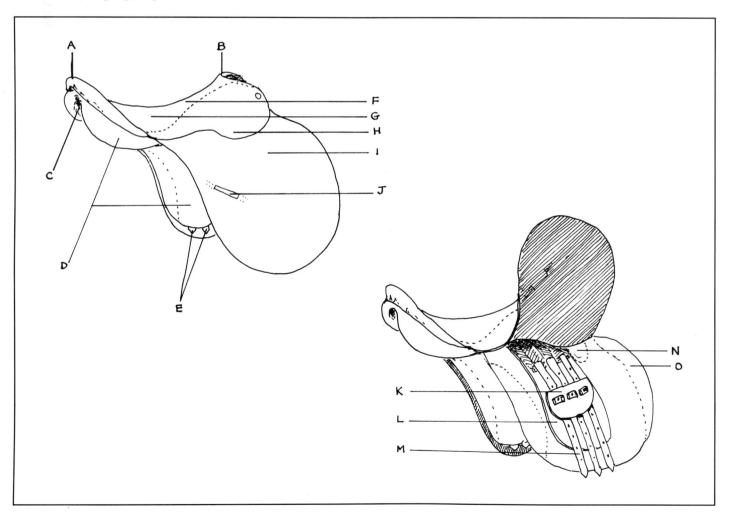

SADDLE FITTING

Saddle fitting can be tricky; you need expert advice if you do not have much knowledge of it, and should also try to develop a 'seeing eye' regarding the balance of a saddle, so you know whether the one you are looking at really does sit level from front to back and side to side on your horse's back. The muscle development of the back can greatly affect saddle fit, and although a good saddler can do a lot to rectify fit by means of skilful stuffing of the saddle panel (the stuffed part under the seat), saddles are now coming on to the market with adjustable trees which can be altered as your horse either develops muscle and gets fitter or, conversely, lets down and becomes fat, or loses condition (weight).

Remember that even on the most superbly-fitting saddle a bad rider can give her horse a sore back and herself a sore bottom. Always sit as still and balanced and relaxed as you can, and try to move in harmony with your horse, not irritating him with excessive movements which go 'against his grain'. This topic is fully covered in Part 7.

The main point as regards fit is to check that when the horse's heaviest rider is mounted no part of the saddle should touch the horse's withers or spine, even when he or she leans forward or back:

there should be a clear tunnel of daylight down his back under the saddle (down its gullet), and you should be able to fit the width of four fingers between the pommel and the withers, and between the cantle and the spine.

The saddle should not be too narrow or it will pinch the sides of the withers; and if too wide it will rock from side to side. You should just be able to slide the flat of the top halves of your fingers around the withers, no more or less. In length, the pommel should go over the withers just behind the highest point. At the back it should *not* extend back to the loins or kidney area (between the back proper and the rump or hindquarters) or it could cause bruising and injury, not least to the kidneys themselves. For similar reasons, never slop around on the back (cantle) of the saddle, but ride properly in the centre of the seat.

The panels of the saddle must not hamper the horse's shoulder movement: this is tested by picking up a foreleg and holding it out straight in front at a comfortable height, such as the point to which the horse would raise it in movement (not exaggeratedly high, in other words). You must be able to place your hand easily between the top of his shoulder blade and the front of the saddle.

The saddle also needs to fit you! You should be able to fit just the width of your hand in front of your body to the pommel and behind to the cantle. When your stirrups are a comfortable length, your knee must not be pushed off the flap or very near the edge, nor must there be several inches of space between your kneecap and the edge of the flap.

When trying a saddle, ride round in it for about 15 minutes to let it and you settle, then check the fit again. Remember, you cannot compensate for a badly-fitting saddle by using a numnah (saddle-shaped cloth or pad) beneath it, thick or thin; the only remedy is to get it adjusted or to try another saddle. More horses than I care to think about have been ruined by back pain due to badly fitting saddles – equalled only, possibly, by badly-handled bits. Horses not only react to pain and discomfort by playing up or napping, but they defend themselves by moving wrongly (as we might in ill-fitting shoes). This causes unnatural stresses and strains on other muscles and ligaments, further pain and injury, and so it goes on.

Do therefore get expert help from your adviser or instructor, as well as a good saddler on saddle fitting: it is vitally important.

GIRTHS

To keep your saddle on you will obviously need a girth, and it is a good idea to have a spare as girths get a lot of wear and tear. Good materials for girths are leather, lampwick and mohair. Leather is not absorbent, however, and many owners do prefer girths which are, as they feel this adds to the horse's comfort (girths get very sweaty, naturally). Lampwick and mohair are both absorbent and I personally prefer them. You can also get cotton string girths which are popular for summer, although there is a slight chance of the horse's being pinched between the strands – and the nylon string girths are, I think, an abomination. They do not absorb any sweat and, in my experience, can be very harsh on the thin skin of the horse's girth area. Some synthetic girths seem quite good and, as fabrics improve, girths are becoming available in 'breathable' synthetic fabrics which permit sweat to evaporate through the girth to some extent.

Any girth, numnah or saddle can make a horse sore if it is not kept clean and soft, however, and this is a most important point in tack care.

Girths (and saddles) come in various sizes, so tell your saddler how big your animal is so he can make an informed estimate of the size you'll need. Full-length girths can be bought, and also short belly girths which fasten below the bottom edge of the saddle panel on long girth straps, ostensibly with the idea of removing an uncomfortable lump of buckles from under the rider's leg. However, with modern saddle designs this drawback does not seem so common and many British-made dressage saddles now do not have the long girth straps which were once such a feature of this style.

Some girths, too, are tapered where they go behind the horse's elbow, giving him more freedom of movement and lessening the likelihood of the skin creasing and being rubbed in this area. Such girths come in two main styles, the Atherstone and the Balding.

SMOOTHING OUT THE WRINKLES

After tacking up, always pull the horse's forelegs forward from the knee, to smooth out and bring forward the skin underneath the girth, for increased comfort and to lessen the possibility of pinching and galling. Bend the leg forward at the knee by clasping your hands behind the knee and lifting the leg. To pull the leg straight up and forward from the fetlock as is often tried is difficult and un-pleasant for the horse, and the method described is just as effective.

Learn to adjust your girth when mounted without looking down as this is safer. Bring your leg forward and hold the flap with your arm. Pull up, or let down, the girth, feeling the holes and buckle guide with your finger. Keep a light contact on the reins with your other hand: *never* let go of the reins.

A selection of girths, from left a lampwick girth, a cotton string girth, an Atherstone girth which tapers in at the elbow to help avoid pinching, a Balding girth of crossed leather designed on the same principle, a brushed nylon girth, a nylon string girth and a short belly girth for use on saddles with long girth straps. The girths are displayed on a woollen Witney-style day rug, so called because of the distinctive striping pattern on such rugs and blankets.

STIRRUPS

Stirrups should be of stainless steel (as should bits) for sufficient strength and, therefore, safety. They often take much of your weight (for instance when jumping), and the nickel ones bend easily or may even snap at a crucial moment. There are many styles, but a basic ordinary stirrup is hard to beat – its object, after all, is simply to take the weight of your leg on the ball of your foot and make riding more comfortable for you and your horse. Specially angled and shaped stirrups can be too restricting and if wrong for you, can even cause pins and needles and cramp. Not exactly conducive to good riding. There is also a safety stirrup, where the outer side is not metal, but a loop of strong rubber so that if you do fall, your foot will press hard against the rubber which will then be pulled off its retaining hook, making more or less impossible the chance of your getting a foot stuck in the stirrup and being dragged to your death (no exaggeration).

You can get little rubber treads to fit in your metal stirrups which make it easier to keep your feet in the stirrup, and are slightly warmer to the feet (so they say) on bitter winter days.

Your stirrup should be an inch (2.5cm) wider than the widest part of the sole of your riding boot. This will greatly help lessen the chance of your foot getting jammed in the stirrup in a fall, or of you getting your foot actually through the stirrup with a similar result.

STIRRUP LEATHERS

At the top of each stirrup is a slit called the eye; pass the stirrup leather through this, then buckle it into a hole appropriate to your length of leg, making a loop, and slot it over the stirrup bar on the saddle panel. (It's safest always to leave the safety catch on the end down, so the leather will slide off in an emergency.) The eye should be just wide enough to allow your leather through without scraping it on either side, which make adjustments difficult and can spoil the leather. Stirrup

To adjust your stirrups when mounted, bring your leg back, keeping your foot in the stirrup. Hold the reins with one hand and feel the holes and buckle tongue with your finger. Make sure the buckle is right up against the stirrup bar when you have finished by pulling down hard on the part of the leather which lies against the saddle flap.

To run the stirrups up their leathers, hold the top piece of the loop formed by the leather up in one hand and simply push up to the stirrup bar, then thread the leather down through the stirrup.

leathers need to be extremely strong, and my favourites are unbreakable rawhide or buffalo hide (check with your saddler). Get a suitable width for your make and shape and stirrup size which is obviously governed by your foot size. If they are too narrow they'll be uncomfortable, and if too wide they'll look silly.

NUMNAHS

As already mentioned, numnahs are saddle-shaped pads used under saddles – *not* to rectify the fit, but to give extra back comfort to the horse (as most of them are at least slightly padded); they also help absorb sweat in an area which becomes very sweaty during work and which is subject to considerable pressure, from the weight of the rider on the back. Especially on long rides, numnahs do increase the horse's comfort and are a useful, if non-essential, addition to his wardrobe.

Probably the most useful are quilted cotton, quite thin but nonetheless effective, and easy to care for as you just wash them – like any cotton garment – in hand-hot water (because the padding is often synthetic) and a mild washing powder or soap (use non-biological, in case of skin reactions). Nylon numnahs are not good, in my experience; they slip in use, they do not create an absorbent medium between leather and back, and can actually cause soreness. Some numnahs are padded and made of synthetic (usually acrylic) fleece: these are quite good. There are also real sheepskin numnahs suitable for winter use (these days they are often treated so you can wash them in tepid water and soap, dressing the leather side with a leather conditioner); sheared sheepskin is also available for use in warmer weather. Various special numnahs which 'breathe' and both spread and lessen pressure and friction are used particularly in endurance riding.

It's all a matter of taste, but I feel numnahs should be unobtrusive and of a colour close to the saddle (brown or black) or a dark, discreet colour such as navy, dark green; or, *clean* white or beige; but *not* bright red or tartan.

BRIDLES

Basically there are bridles meant to take one bit, normally snaffles; bridles to take two bits (double bridles); and bitless bridles, which are something of a specialist area.

Most people start off with a snaffle bridle, and the parts of such a bridle are shown in the accompanying drawing. A double bridle will have an extra headstall or a piece of leather with a shorter, separate cheekpiece which buckles to it for adjustment of fit, to take the curb bit. The bridle should fit as shown in the accompanying photographs, and the line drawings show how to put it on and take it off. Good fit is essential, as badly fitting bridles are a cause of head-shaking and head-tossing, resistance and lack of co-operation, and this is not only unsafe but unkind to the horse. Bridles meant to take a snaffle bit can also take a pelham bit. Only experienced riders should use double bridles as they require considerable skill to handle initially, and to use them properly takes practice over a considerable period of time. Basically, it is not suitable simply to add an extra headstall to a snaffle bridle to convert it to a double, as the bridle leather on a snaffle is normally wider and that is what it will always look like – a 'converted' snaffle bridle. Double bridles have narrower leather to avoid giving the horse's head a cluttered look.

Bitless bridles can be quite severe. There are different sorts. They are useful for horses with mouth problems (perhaps due to previous rough riding) and those having temporary trouble with a mouth injury or teeth so that work can continue. A useful design for novice riders is the Scawbrig which has a padded nosepiece with side rings through which pass reins attached to another padded section in the chin groove. A feel on the reins creates all-round pressure on the nose bone and jaw, giving good control.

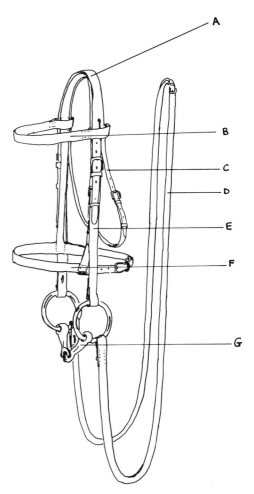

Parts of a snaffle bridle:

A   Headpiece
B   Browband
C   Throatlatch (pronounced 'throatlash')
D   Reins
E   Cheekpiece
F   Noseband (in this case a simple cavesson)
G   Bit (in this case a loose ring, jointed snaffle)

PUTTING ON A BRIDLE

To put on a bridle, first put the reins over the horse's head and lay them just behind the ears, like this, so you can catch hold of them easily should he move away. Hold the bridle by the headpiece with your right hand and support the bit in your left, like this. The throatlatch and noseband must be undone.

For a slightly difficult horse, you can, instead, hold both cheekpieces together in your right hand with your arm under his throat instead of over his poll, so you can control his head with that hand. With either method, position the bit exactly where the teeth meet (not up or down on the gums which is useless and may hurt him) and press gently when he should open his mouth. If he doesn't, put your thumb in the corner of his mouth like this (there are no teeth there as it is where the bit rests, so you won't get bitten) and tickle his tongue. This almost always results in his opening his mouth. Then carefully and quickly slip in the bit.

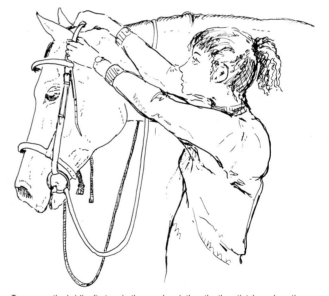

Bring the bridle upwards with your right hand, obviously drawing the bit up with it, and gently and quickly put the headpiece over one ear and then the other. Carefully straighten out the mane hair under it, and bring the forelock out *over* the browband. Check that the browband is level and also the bit. If the latter is higher at one side than the other, slide the headpiece carefully to right or left at the poll, as necessary and/or adjust it at the cheekpiece buckles, using the same holes each side. If the noseband is crooked, slide its headpiece (which goes under the headpiece and through the browband's end loops) from side to side at the poll, as needed, lifting the low side first, to get it straight, as this is more comfortable for the horse. Fasten first the throatlatch, then the noseband.

To remove the bridle, first undo the noseband, then the throatlatch, and gently ease the headpiece up and over the ears from the back with both hands, one at each ear. Allow the horse to let go of the bit in his own time. If he hangs on to it, tickle him inside the corner of his mouth to encourage him to let go, but *never* pull the bit out of his mouth as this can really hurt and is often a cause of horses becoming difficult about the head (called 'head shy'). You can bring the reins over the ears at the same time as the headpiece.

This horse is wearing a flash noseband. It's like an ordinary cavesson with a section going round below the bit in the chin groove. Fastened snugly but comfortably, it discourages opening the mouth and evading the bit, providing more control.

This is a drop noseband. The front piece should come *no lower* than this. The back straps pass under the bit in the chin groove. Again, the aim is to discourage the horse from opening his mouth and evading the bit.

FITTING A BRIDLE

You must be able to fit the width of your hand into the throatlatch when fastened.

The browband should be short enough to keep the headpiece up behind the ears without pressing it into them, and you should be able to slide a finger easily underneath it.

You should be able to fit your fingers up under the noseband as illustrated. This cavesson noseband is the type having slots to accommodate the headstall rather than its being stitched on: this balances the noseband better on the head. The noseband should come about half way between the corners of the lips and the sharp face bones. The bit, in this case a loose ring snaffle, should *just* wrinkle the corners of the lips.

BITS

A selection of fairly common bits is shown in the accompanying photograph. Basically, most well-trained horses go well in a single jointed snaffle bit (some have two joints with varying actions). Some snaffles, however, have a half-moon shaped (mullen) mouthpiece, and although these are said to be mild many horses lean on them and become unresponsive, and even hard-mouthed as a result. A jointed bit (sometimes, though fairly rarely, called a 'broken-mouthed' bit because of the break or joint) acts mainly on the bars of the mouth, that toothless area of the gums just where the corners of the mouth are (a physical feature which is highly convenient for us) with the tongue fitting into the space provided by the joint. A mullen-mouthed bit, however, acts mainly on the tongue and, as the mouthpiece is stiff as opposed to moveable, the horse can manipulate it and lift it up with his tongue to prevent bar pressure. Depending on the horse and his preferences, this may be a good or bad thing. It is essential to discover through trial and error, with expert supervision, in which sort of bit your horse goes best and most comfortably (preferably the mildest consistent with control) and stick to that.

You can, in fact, get snaffles with a dead straight bar mouthpiece but most horses are uncomfortable with the tongue-squashing effect these can have, although some find rubber ones acceptable.

The way the rings are attached to the bit also has a bearing on how the bit feels to the horse. The mouthpiece can be joined to the bit rings by an extension of itself up and down the ring. As this is slightly egg-shaped, such bits are called eggbutt (butt meaning end – of mouthpiece) bits. Alternatively, the mouthpiece can have holes in each end through which the bit rings slide. These are called 'loose ring' bits, and do indeed give a looser feel which many horses prefer to the more fixed feel of the eggbutt, but unless great care is taken over fit it is possible, particularly if the holes are worn with use, for the corners of the mouth to be pinched extremely painfully between the mouthpiece and the ring, in the hole. This is impossible with the eggbutt. In the 'loose ring' bit the rings are flat; thinner, round rings have the name 'wire ring' and are normally larger than the loose ring ones – they are popular in racing as they are harder to pull through the horse's mouth, which can often happen in situations where control is, shall we say, lacking.

Some snaffles have a metal extension at right angles to each end of the mouthpiece to make pulling through impossible, such as the Fulmer snaffle illustrated, but these are not as universally popular as they were a generation ago. They are never used in racing in case they pierce the horse's cheek in a fall, a frequent occurrence in that sport, but quite likely in others, too They were originally used for young horses to help them understand turning aids (and are commonly used for this at the Spanish Riding School in Vienna), and were never designed for fast work.

Pelham bits are an attempt to give the rider the stronger leverage of the double bridle curb bit and the horse the non-jointed snaffle, all combined in one mouthpiece. In a double bridle the object of the curb bit (also called a Weymouth) is, in combination with the curb chain which lies in the chin groove, to provide leverage but also to encourage the horse to relax his jaw to a light touch and flex at the poll. The rider has two sets of reins, one for the bridoon (a thinner version of an ordinary *jointed* snaffle) and one for the curb, the curb reins being fitted to the bottom of the curb cheek resulting in an indirect, leverage effect. In a pelham, you get the leverage *and* the (non-jointed) snaffle effect, as the snaffle reins attach to rings on the end of the pelham's mouthpiece. However, for those finding two sets of reins too much to handle, leather loops called 'roundings' can be attached to both snaffle and curb rings and the reins attached to the rounding; however, you then get a less precise effect from the single rein. This is a popular device for rather strong children's ponies.

A well-fitting bridle with brass browband, padded cavesson noseband and eggbutt snaffle bit.

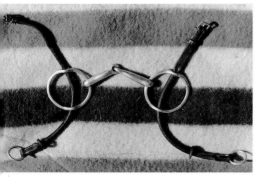

A gag snaffle. The buckles fasten to the top halves of the bridle cheekpieces and the rings take the ends of the reins. To be used correctly, a pair of reins should be fitted to the bit rings in the normal way, the gag reins only being used when the horse gets his head down or starts to 'get away'. Novice riders should not be riding horses who need gags.

Many people say pelhams are neither one thing nor the other, and they are right; but the fact remains that many animals go well and happily in them, and that is worth a lot.

Bitting is a vast subject, and while you will probably be advised to start off with some kind of ordinary snaffle, it pays to learn as much as you can on the subject. One piece of wisdom which is well worth remembering, however, is this: it is not so much the bit in the horse's mouth which matters, as the sensitivity and skill of the hands on the other end of the reins – yours!

Bits come in stainless steel, nickel (not advised, for the reasons already given, see p.124), vulcanite (hardened rubber), rubber (very mild and soft) and nowadays various plastic and nylon materials, too, which are very light and seem comfortable. Generally, you won't go wrong with stainless steel, although some animals do prefer a softer feel.

As for width, the bit should protrude just a quarter of an inch (about .75cm) on each side of the horse's mouth: if wider it will slide about and bruise his mouth, and if narrower it will pinch him.

A selection of bits, from bottom left clockwise, a wire ring German snaffle (thick mouthpiece), an ordinary eggbutt snaffle, a Fulmer snaffle, a vulcanite mullen-mouthed (half moon) pelham with roundings fitted, a port-mouthed snaffle to discourage a horse from putting its tongue over the bit, a rubber, eggbutt, mullen-mouthed snaffle and a roller snaffle to discourage horses from pulling and leaning on the bit.

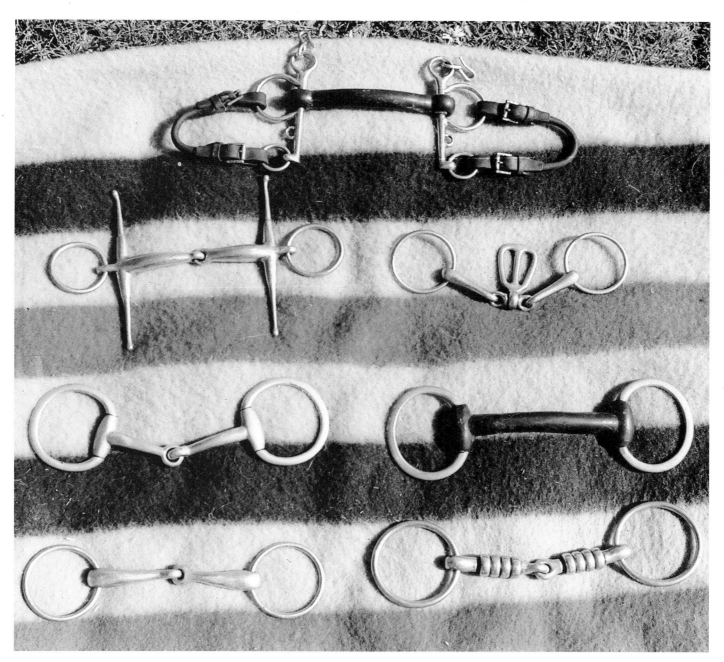

REINS

For general riding, probably the best sort are rubber-covered leather reins which will give security of grip even when wet with rain or sweat, and will not stretch in use as can plaited leather. Plain leather are only alright on well-mannered horses in dry weather. Rubber does wear out, though, and to have reins re-covered can weaken them. If this proves too uneconomical for you, consider laced leather reins, with a herringbone pattern of leather lacing on plain leather. You can have plaited leather or cotton but, as stated, these can stretch in use. Plaited nylon reins look cheap and are rough on the hands.

Try the reins in your hands to make sure they are a comfortable width. If they are too wide you'll be uncomfortable and could prove clumsy handling them, but if too narrow they encourage you to clench your fists into an unfeeling, insensitive ball which won't make for harmonious communication with your horse via the bit.

Reins are sold in pairs. One end attaches to the bit ring, normally with a hookstud fastening which the saddler will show you how to operate, and they buckle together at the 'rider' end. They come in different lengths, too, shorter for jumping, longer for general riding and the larger animals. Again, tell the saddler the size of your horse and that you want the reins for general riding, and he'll help you get the right length.

HEADCOLLARS

A headcollar is another vital piece of equipment. It is rather like a heavy-duty bridle with no bit, but of course has no reins either and many have no browband. I do prefer a browband as this prevents the headpiece sliding down the neck and creating an uncomfortable pull on the front of the nose, particularly for horses wearing their headcollars for long periods.

The best type have browbands and adjustment buckles not only on the nearside cheekpiece, but also on the noseband and often on the throatlatch. Basic headcollars have only the buckle on the cheekpiece. You undo the buckle to put the headcollar on, stand with your back to the horse's tail, hold the headcollar by both cheekpieces and bring it up over his muzzle. You then flick the headpiece over his poll with your right hand and fasten it. To remove it, undo the buckle and just slide it down.

Your leadrope, which is separate, will probably have a dog-lead type of trigger clip (the safest) or a spring slip. Either way, it is safest to have the fastening facing away from the horse's head, towards his neck.

If you have a choice of leadrope, get the longest you can and tie a knot in the end, making it more difficult for it to be pulled through your hands should the horse become startled and take a pull. Always wear gloves when leading your horse for any distance, such as to his field, as they will protect your hands and prevent ropeburns, enabling you to keep hold in an emergency, hopefully!

Headcollars come in leather and very strong nylon with, occasionally, fancy rope and also cotton webbing. Leather will probably break should the horse become caught up by his headcollar in the field and take a pull, whereas nylon will not, so although nylon ones are comparatively cheap, and fine for general stableyard use, if your horse needs a headcollar on when turned out (perhaps because he is hard to catch) *don't* make it a nylon one.

MARTINGALES

There are two main kinds of martingale: running martingales and standing martingales. At one time standing martingales were very common, but today they are mostly seen on the polo field, with some exceptions.

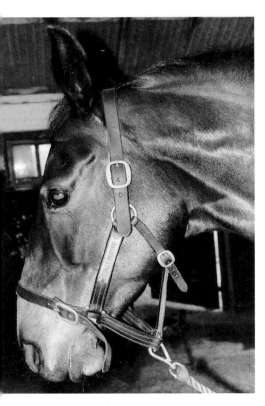

A well-fitting, well-kept leather headcollar with, unlike many, the advantage of being able to be adjusted on the throatlatch and noseband as well as the headpiece. The noseband could possibly be a little higher and tighter, and an extra refinement would be a browband which stops the headpiece sliding down the neck and causing a pull on the nose when the headcollar is worn for long periods.

A running martingale can, in fact, be useful for a novice rider not yet in full control of her hand position, and is used as such by some riding centres. It consists of two leather straps: one encircles the base of the horse's neck, the other passes at right angles through the bottom of the circle, one end running down the breast between the front legs and fastening round the girth, the other end dividing into two further straps, each ending in a ring through which one rein passes. The correct adjustment is that the rings only start to pull on the reins when the horse lifts his head too high or the rider does the same with her hands. The result is the same – the martingale lessens the effect of the incorrect head or hand position on the bit in the horse's mouth by keeping the rein action more nearly correct.

A standing martingale fastens with one strap (in place of the running martingale's two) on the back of a cavesson noseband and is supposed to discourage, or prevent, the horse from throwing up his head (usually against the action of the bit) beyond the point of control. When the horse's head is held in a correct, normal position it has no effect if adjusted correctly, but only begins to pull on the noseband if the head goes too high. In practice however, these martingales can encourage horses to lean on the noseband and develop muscles in the neck which enable them to pull hard, and in this way can be counterproductive.

If a horse is felt to need a martingale at all, discuss matters with your instructor first. A running martingale may help an inexperienced rider, but standing martingales should generally be done without.

A third little martingale which can be useful with horses who toss their heads about (and you should find out why they are doing this) is the Irish – simply a short length of leather with a ring at each end through which the reins pass, the martingale lying between the horse's chin and neck. Horses who habitually toss their heads are not suitable for novice riders, so you should not need this little item, which does prevent the reins on such horses being tossed over to one side of the neck, greatly reducing your control.

There are several other martingales or schooling aids, none of which I have the space to describe here and none of which are needed by inexperienced riders. As you progress, however, it is useful to know about them and any competent instructor, or more advanced book, can give you full details.

BANDAGES AND BOOTS

Again, the selection here can be quite bewildering so it may be good to know you *can* probably manage without them.

Stable bandages are best made of wool (the soft, 'mouldable' type rather than the stiffer, more expensive velour type) or an acrylic knitted fabric, and are good for providing extra warmth on a cold night. They are also good for drying off wet legs, and are used over padding to cushion the horse's legs against possible knocks.

Bandages for work, called exercise bandages normally, are best made of crêpe which has some stretch in it and must always be put on over padding, either proprietary sheets of padding which you cut to size, Gamgee Tissue – cotton wool covered with gauze to stop it breaking up – foam rubber or plastic (not good as it compresses and doesn't even out the pressure of the bandaging as it should), thin felt fabric or cotton wool. The first two are best in practice.

The accompanying drawings and photographs show how they should be put on, and how they should look when finished. Bandaging can do immense harm if wrongly done and you really should practise well on a table leg or young tree trunk or fence post before trying your skills on a horse (who may not stand as still, either). It also takes considerable know-how to judge the tension, so it's well worth paying your instructor for a proper lesson. You may need to know how to do it to dress a wound, anyway, so it is something you should learn.

A horse wearing a running martingale. The rider's hands are a little high and the martingale is just coming into action on the reins so that she does not create an upward feel on the bit.

A standing martingale just coming into action on the cavesson noseband as the horse raises his head a little high. Such martingales must never be fitted to drop nosebands, grakle nosebands or any part of a composite noseband (such as a flash noseband) which comes below the bit as the action would then be much too severe, but always to a cavesson noseband proper or the cavesson part of a composite noseband.

BANDAGES

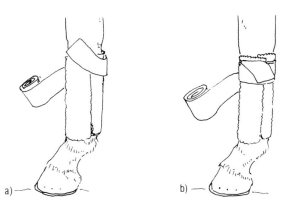 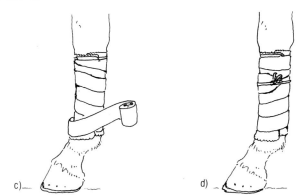

a) To apply an exercise or work bandage, first damp the hair slightly which helps 'hold' the padding, then apply your padding with a double thickness over the tendons. Start the bandage like this leaving a short end upwards . . .

b) . . . then take one turn round the leg and drop down the spare end, which will help 'lock' the bandage in place. Take a turn over it . . .

c) . . . and continue bandaging evenly down the leg, not stretching the bandage too much. Stay on the padding and turn back up the leg.

d) Try to finish the bandage with the end pointing backwards (so it is less likely to get caught in twigs etc.) and fasten the bow between the leg bone and the tendons, tucking in the spare ends for neatness and so that they do not get caught and possibly pull the bandage undone, when it might trip the horse.

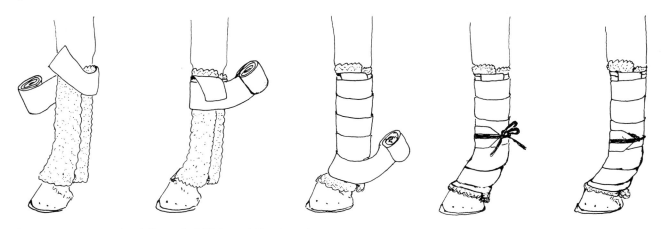

Stable bandages are put on in a similar way, if slightly looser, but bandage and padding go right down over the fetlock. At the fetlock, you'll find the bandage takes a natural turn upwards. Again, finish the tie as shown and remember to tuck in the loose ends.

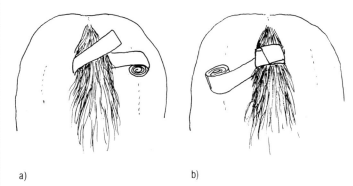 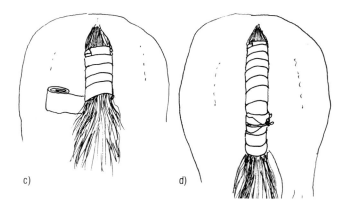

a) Tail bandages must be very carefully put on to ensure that they are *not* too tight (as no padding is normally used), yet firm enough to stay on, and there there are no wrinkles in the bandage to cause pressure or soreness. Start off like this right up at the root of the tail.

b) Take one turn round . . .

c) . . . and let the spare end drop down. Turn over it with the bandage and continue evenly on down the tail.

d) Go about three quarters of way down the dock (some people go all the way) and turn up again. Finish off with the knot on the front-side so that should the horse lean on his tail, particularly when travelling, he will not cause a pressure sore by leaning on the knot.

To remove a tail bandage, just grasp it at the top with both hands and slide it off down the tail, making no effort to undo it.

Exercise bandages do not, as is popularly supposed, support tendons (they would need to restrict the action of a related joint to do that) but they can lessen concussion and protect a horse's legs from knocks, cuts and thorns during active work.

The main points to remember with all bandages are that the pressure and tension must be even all the way, and the fastenings must be the same tension as the bandage. Be particularly careful with crêpe bandages, which have a self-tightening effect – do not stretch them unduly when putting them on, otherwise they could become critically tight on your horse's legs and seriously lame him. I have seen racehorses' legs burst right open down to the tendons through this, and such injuries take months to heal – if everything progresses normally. Should infection set in, things can get very serious indeed and take a long time to get right, if at all.

If horses kick themselves when in motion either through faulty action, lack of fitness, or tight turns and awkward movements in fast work (which can afflict any horse), they are best worked in protective boots. You can use brushing boots for such horses when they 'brush', which usually involves self-inflicted wounds low down on the inside of the legs; or speedicut boots – the same, but they reach higher up the leg – such horses being said 'to speedicut' but don't ask me the origins of the word! These boots have reinforced pads (leather or synthetic) over the inside of the fetlock and up the inside of the leg, front or back, and fasten with straps and buckles or clips on the outside of the leg, with the opening of the straps facing backwards so the horse is not so likely to catch them and force them open. Sometimes the straps are on an elasticated base strap for extra comfort and more even fit. Velcro fastenings are very often used, too. You can put them on over padding if you wish, although this is not strictly necessary. One disadvantage is that grit can work down between the boot and the leg and cause soreness, and in some cases, swelling. However, not all horses are so afflicted and boots are certainly quicker and easier to apply than bandages. They should just be comfortably snug so as to stay on and in place (no twisting round) without causing discomfort.

Over-reach boots are like rubber bells and are used for horses who over-reach (tread on their front heels with their back hooves). This is common in young horses or during fast work in athletic performance animals. The best and easiest type of over-reach boot to use, and the one which tries the temper less, is the one which opens at a fastening round the pastern area so you aren't struggling to pull them on over the hooves (not as easy as your instructor makes it look!) One type is made of overlapping synthetic 'leaves' which also obviate the common problem of the boot turning inside out (like a little basin round the fetlock) during work and becoming useless and irritating.

RUGS

There have been great advances in horse clothing over the last twenty years, and especially the last ten. From the old woollen day rugs and jute night rugs we have moved to permeable ('breatheable') fabrics which let moisture, whether rain, tap-water or sweat, evaporate up through the fabric yet can stop water from outside getting in. Even if the fabric is not completely waterproof, it may well be showerproof, and these permeable rugs do help greatly with horses who sweat more than you'd want under their clothing. Some do it from excitement and anticipation, others habitually break out in a sweat after work (although this is mainly due to insufficient cooling down after work) and permeable rugs do help these horses. In any case, they help maintain a healthy body atmosphere, for want of a better expression.

There are also other, non-permeable synthetic fabrics, both acrylic and nylon which is usually quilted and with a brushed cotton lining, which all have the

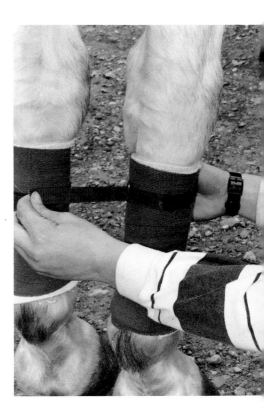

Exercise bandages can be tied on by their tapes, secured with adhesive tape of various sorts, stitched on (super-safe for fast work) or quickly secured with Velcro fastening, as here. The longest length of tape is passed through a small metal loop on the end of the short length which passes the other way round the leg. The long length is then brought back on to itself and pressed down on to the velcro. Whatever method of fastening is used, the tape/s should be the *same* tightness as the bandage itself. If too tight an injury could be caused and if too loose the bandage may unravel.

To remove bandages, simply undo the fastening and unwind them round the leg, passing them quickly from hand to hand in a bundle (without re-rolling). Rub the legs briskly for a few seconds. Re-roll bandages by folding the tapes across the top to form a 'core' inside as you hold the bandage up and roll it down on itself towards you without stretching it. New bandages are always rolled the wrong way with the tapes outside and you have to roll them properly before being able to use them.

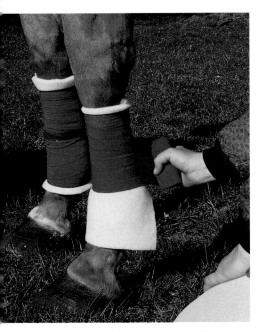

All bandages, stable and exercise, should be put on over padding. Specially made padding material is shown here, available from saddlers, but cotton wool or, better, Gamgee Tissue, can also be used. The padding must cover the whole length to be bandaged.

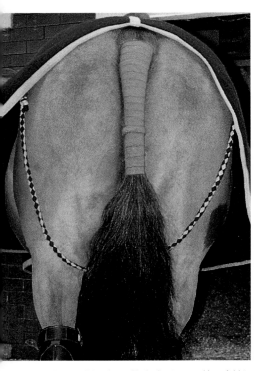

A neat tail bandage with the knot covered by a fold to prevent it being rubbed undone. The rug's fillet string is the correct height and *under* the tail.

advantage of being easy to launder and quick to dry, unlike traditional materials which, apart from the cotton and linen still often used for summer sheets (lightweight summer rugs), are heavy, space-consuming and can take days to dry.

The shape and fitting of the modern rugs has also greatly improved. Although many people do still use the old-fashioned and inevitably somewhat uncomfortable (for the horse, obviously) non-shaped rugs with surcingle or roller fastening, the modern rugs are shaped according to the horse's body, and fasten with various types of thinner surcingles which criss-cross loosely under the horse's belly or form an actual harness to keep rugs on without those relatively tight, 'belt'-type surcingles; thankfully the modern style is becoming more and more common, to the benefit of the wearers.

What do we need rugs for? What jobs should a good rug do? What are the disadvantages of a poorly designed, badly fitting rug?

Rugs are used, perhaps along with blankets or under-rugs, to provide warmth to a clipped horse in winter. Horses are clipped to enable them to work hard in winter without sweating up too much or chilling as they cool down in long, wet coats. Removing the coat means we must make up for this when the horse is not working – in stable or field – by providing clothing. Sometimes clothing is worn during work in the form of a sort of half rug, or rug without a front. These are called exercise sheets and there are various sorts, waterproof and otherwise. Rugs for use on outdoor horses in the field must also be wind and waterproof.

Traditional fabrics are wool for smart day rugs, with cotton bindings in a toning or contrasting colour and often bearing the owner's initials, wool (striped or otherwise) for blankets to use under rugs, jute or sailcloth sometimes lined with wool for night rugs (rougher affairs so it doesn't matter if the horse lies in his droppings), and heavyweight waterproof canvas, usually half lined with wool, for turn-out rugs.

Modern, synthetic fabrics are very varied and often have patented brand names such as 'Thermatextron' which tell you nothing – and if you ask you are unlikely to get an answer unless you make your questions quite pointed! Seriously, most of them are nylon, polyester or acrylic based, although there are others; the main qualities are lightness and ease of laundering plus excellent wearing qualities. As mentioned, many allow evaporation to leave the horse through the fabric and are called permeable or 'breathable'. Some are lined with polyester quilting, some are filled with polystyrene beads and some are of special heat-retaining material for extra warmth without additional under-rugs or heavy fabric. Some have special under-rugs which clip into the top rug or stay in place by means of very pliable fabric and good design, for use in very cold weather.

Anti-sweat rugs are often seen, and can be very useful in drying off wet or damp horses. Contrary to popular opinion, they do not stop horses breaking out (in a sweat) after work or a hard day or prevent their doing so from excitement. As they help horses dry off quicker than rugs of traditional fabrics it only *seems* that they do these things.

What actually happens is that when the rug is worn alone, the string-vest type of fabric from which it is made creates more little eddies of air disturbance over the skin than would normally occur, and these dry off the moisture by encouraging it to evaporate, and by wafting it away, quicker than if the horse wore nothing or wore ordinary rugs.

When the rugs are worn, as they were originally intended, next to the horse with another rug or sheet on top, again a significant layer of air is trapped next to the horse and insulates it, keeping it warm in cold weather and helping moisture evaporate, albeit more slowly than if the rug were worn alone. It is not a good thing in chilly weather to cool the horse down *too* quickly (this can lead to chill and cramp), but in hot weather it may be desirable.

These rugs are also good to use under an old rug, in place of thatching the horse with straw to hasten drying off, as many yards do not use straw now.

PUTTING ON A RUG

The accompanying illustrations show one way of putting on a rug. Fold it so the back is folded over to the front with the lining showing. Put it over the horse's withers, well forward and still folded.

Hold the front firmly down and fold the back of the rug over the quarters. Then slide it back *slightly* into place, smoothing the hair comfortably back. Fasten the surcingles first, then the breast-strap.

With rugs having crossed belly surcingles, the right front surcingle comes under the belly and fastens to the left back, and the left front to the right back. They should almost touch the belly underneath. If you can fit the width of your hand between surcingles and horse, like this, they are about right.

The lower photograph shows a good quality, well fitting rug of comfortable, modern design, but the breast strap is fastened too tightly causing a pull on the points of the shoulders.

To remove a rug, unfasten the breast strap first, then the surcingles (which can be clipped up to themselves to keep them from being trodden on). Hold the rug front and back on the back seam and just slide it off over the tail. This way the lining is not exposed to any bits which may become embedded in it and either rub the horse later or be difficult to remove completely. Fold the rug down the middle into a square.

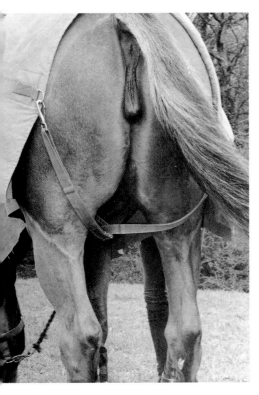

New Zealand rug leg straps linked through each other correctly and adjusted to the correct length.

A well-used, well-fitting New Zealand rug of comfortable, modern design. It comes forward of the withers and allows the pony to graze without pulling on the neck or shoulders, it extends well back just past the root of the tail, it has no surcingle, being properly shaped and is long enough to protect the belly from wind.

Turn-out rugs, commonly called New Zealand rugs, are now of various designs and, again, of lighter but waterproof materials with fillings or linings. They have legstraps or special harnesses to keep them on and some have tail flaps and related neck or head-and-neck hoods for extra protection.

The range now is truly vast and you are advised to study advertisements in equestrian magazines, look at your saddler's stock, and send to the manufacturers for full details of any rug you are interested in, as not all retailers can give you full details of the more advanced rugs.

The elements of fit and style are basically the same for any rug whatever its purpose. It should come *forward* of the withers, not rest on top of them, and reach right back to the root of the tail, extending just past it for a turn-out rug. The back line should undulate to fit the shape of the spine of the horse – except in the case of very pliable, soft underblankets and under-rugs, rising for the withers, dipping for the back, rising again for the croup ('top' of the quarters) and down again towards the tail. The good modern rugs also have shaping at the buttocks, shoulder, elbow and stifle, and often all four places, to create proper fit (as human clothes are darted and tucked) and lessen pressure and pull on the points of the shoulder and hip and on the croup.

Most importantly, a good modern rug will not employ a 'round-the-middle' type of surcingle, or a separate roller, to keep it on. The shaping does half the job, and harnesses or crossing belly surcingles do the rest. By fastening diagonally under the horse from front to back, perhaps linking at the sides, too, the surcingles counteract movement and pull and normally keep the rug in position very well. The whole point is that such surcingles should *not* be at all tight, and from the point of view of the horse's comfort this is their main advantage; the rugs will fasten at the breast in the usual way. Turnout rugs may simply have a breast fastening, and leg straps which link through each other between the hind legs, the left strap fastening round the left hind leg and the right passing through it and fastening round the right hind leg. This holds them away from the sensitive skin inside the horse's legs and, again, counteracts sliding about. The horse can lie down, roll, play and gallop about and his rug will stay on and fairly well in position provided it is correctly shaped, fitted and adjusted.

Traditional rugs, too, must not be cut in a straight line down the back; there is no need for it, so avoid rugs cut like this – if we all stopped buying them the makers would stop making them. They have to be kept on by means of a conventional surcingle or separate roller, perhaps with an extra pad. This *cannot* be as comfortable for the horse, who, remember, has to lie down and expand in this contraption. An extra plain rectangular blanket may be used underneath for extra warmth, or more unusually a shaped under-rug. The method of putting on both types of rug is shown in the accompanying illustrations.

To rug up a horse with a traditional blanket, rug and roller or surcingle, first lay the blanket on the horse lengthwise so it reaches from tail to ears, and fold up both front corners against the neck, as shown.

Put on the rug and bring back over it the folded front part of the blanket. Fasten the roller or surcingle over the lot, as shown, to help prevent the blanket slipping back under the rug as they so often do, then fasten the breast strap of the rug. Smooth out any wrinkled fabric under the roller and pull rug and blanket forward a little in front of the roller and behind the elbow to take up any slack fabric and give a roomier feel for the shoulders. Do this on both sides, and finally smooth out the mane hair over the withers under the blanket to make the horse comfortable. The roller should not be as tight as a saddle girth as the horse must be able to lie down in comfort. You must be able to fit the flat of your fingers easily between roller and rug and pull it away a little. This method of rugging is obviously not as comfortable for the horse as the modern-style rugs with no roller and will hopefully become obsolete before long.

An exercise sheet fitted under a saddle. The girth (in this case the short belly girth used with long girth-strap saddles) passes through the loop on the bottom edge of the sheet to help keep it in position. Like a numnah, the exercise sheet must be pulled up into the gullet.

All dressed up with somewhere to go. Fully kitted up for travelling, this horse is wearing a headcollar with poll guard, a day rug with roller (with padding underneath), a breastgirth to prevent the roller slipping back, a tail bandage covered by a tail guard tied to the roller to prevent it slipping down, stable bandages over padding, knee pads, hock boots and over-reach boots in front. Some people might add over-reach boots behind as an added precaution.

Side view of the poll guard, showing how the headcollar's headpiece slots through to secure it.

Knee pads should be fastened with the top strap tight enough to keep them in place but no tighter, and the bottom strap very loose, like this, so as not to hamper movement or pull the pads out of place when the leg bends.

Hock boots. The top strap is fastened snugly but not too tightly, to keep the boots in place, and the lower strap is left fairly loose for ease of movement.

CARE AND CLEANING

All leather items, from saddles and bridles to leather straps on rugs or boots, are cleaned in basically the same way: you need a bucket of warm (not hot) water, a washing sponge, saddle soap (the type containing glycerine is generally regarded as best) and a soap sponge which you *never* put in the water. You may also have a chamois leather for drying, a stable rubber and metal polish and duster.

You strip your saddle (remove stirrups, leathers and girth) and firmly wash off all grease, sweat and mud with the washing sponge. If the leather is very dirty put a dash of mild washing up liquid in the water to help break down the grease. Needless to say, cold water is no use for this job. Gently scrape off little hardened lumps of grease ('jockeys') with a coin or blunt knife.

When you have cleaned the leather, take your bar of saddle soap, dip it in the water and rub it on your soap sponge. The sponge must *not* be wet; if the soap lathers it is too wet. Rub the soaped sponge firmly over the leather in small circular movements, paying particular attention to the undersides and working the soap well down into the leather, all over the saddle which can be held on your knee, or on a saddle horse which will turn over so you can get at the underneath easily.

Hang your stirrup leathers by the buckle on a hook, if possible, and wash and soap them by wrapping the sponges round them and rubbing firmly and quickly up and down, paying special attention to the fold in the leather which takes the stirrup and which is most subject to wear. The stirrups can simply be washed or done with metal polish, and polished clean depending on how shiny you want them.

Tack cleaning gear. From left, a thin sponge useful for reins, bridle and leathers, a chunkier sponge easier for cleaning the saddle, soap sponge and bar of glycerine saddle soap on a stable rubber. The empty feed supplement tub at the back can hold water although a bucket is more usual.

If the girth is leather do it the same way. If synthetic or fabric, wash it in warm soapy water (no harsh detergents), scrubbing it gently if necessary; rinse *very* thoroughly and hang it up by the buckles on *both* ends to dry somewhere warm, so it hangs in a U-shape which prevents the water running down on to one set of buckles and possibly rusting or greening them.

Your bridle can be taken to pieces (be sure you know how to put it back together again) and done on a flat surface or hung on a bridle hook as it is, although it is a good idea to move the buckles a hole so you can get at the leather behind them which is subject to most wear. It's also best to undo the hookstuds on the ends of the cheekpieces and reins so you can soap the insides of the turns and keep the studs easily workable. Few things are more infuriating, time-wasting and indicative of sloppily cared-for tack than stiff, stubborn hook studs!

Poke any soap out of the holes with a matchstick, clean the buckles sometimes with metal polish (cleaning it off the leather at once), but *never* do the bit mouthpiece with metal polish. As you go, inspect the tack for worn stitching, fraying, cracked leather, and if you spot any, take the item to a saddler for repair as your life could depend on it.

Ideally, you should clean tack fully after each use, but many owners do not have time. You should at least clean the undersides which touch the horse, and the stress areas such as leather folds and turns. If you use glycerine saddle soap there is little need to oil your tack with neatsfoot oil or proprietary leather dressings if it is in regular use. However, a good dressing can with advantage be given to new tack to help make it more user-friendly, and to little-used tack, or tack going into storage. Put a thin film of dressing on the buckles, too, but never the bit.

Even if you do nothing else, you should rinse your bit thoroughly in clear water after each use so the poor horse is not presented with a revolting bit next time round.

Numnahs, rugs and bandages (which should be put in a pillowcase to stop them tangling) can be washed with mild detergent in a domestic washing machine at an appropriate temperature (usually no more than hand hot as some of the new synthetic rugs with special fillings will melt and go hard otherwise!). Rinse very thoroughly and use a fabric softener if you wish. Remember to look for any manufacturer's washing instructions or ask your supplier if in any doubt, otherwise you could ruin the item and fail to get any recompense because you didn't wash it correctly.

Ideally, woollen day rugs should be cleaned professionally, but you *can* do them yourself (also woollen blankets) in just warm water and washing suds (soap), rinsing very thoroughly in water which is the *same temperature* to avoid shrinking. It is *not* necessary to remove leather straps first, provided you give them a really good oiling before and after washing. Jute night rugs can be washed in the same way, but turn-out rugs made of heavy canvas can just be hosed down on the outside and the lining vacuumed very thoroughly. If you feel the lining is really dirty, you're faced with the laborious job of hand-scrubbing it with warm soapy water, throughly hosing out all soap with warm water, then spreading the rug inside out over a fence or hedge, or hanging it to dry on an old-fashioned and extremely useful clothes rack in a warm room. Obviously, remove leg straps first and clean them as for leather. Oil the breast straps and leg straps regularly as they will be constantly wet and in need of it. (It's obviously best to have two turn-out rugs so you can always have one clean and properly dry and maintained – and the new synthetic ones are so much easier to care for!) In any case, never hang a wet canvas New Zealand rug on the clothes line to dry as it will probably snap it – they're extremely heavy, which gives you some idea of what the poor horse has to put up with when wearing one of these in wet weather, maybe for many hours at a time.

All turnout rugs need periodic re-proofing which can be done with a spray-on product available from saddlers. This should be done anyway after washing.

A cleaned saddle on its horse, with the stirrups run up the leathers.

6 BREEDING
including Veterinary Considerations
by Simon Wolfensohn, MA, Vet MB, MRCVS

B reeding is a specialised subject within the horse world and may seem out of place in a book aimed at not-so-experienced owners and riders or those with a casual interest in horses. However, we frequently come into contact with breeding stock – on hacks, at shows, on neighbouring premises and maybe at livery in the same stable we keep our own horse. Most people like young animals such as puppies, lambs and so on, and don't realise – or just forget – how formidable a fully grown cow, horse or even sheep can be. This section is included to give readers an introductory knowledge of the practical aspects of breeding and the management of breeding stock. A veterinary section has also been added to this part by guest contributor Simon Wolfensohn.

PRACTICAL ASPECTS

What are you aiming for?
Anyone intending to breed a horse or pony should have a clear idea of the sort of animal they wish to produce. They may simply feel, quite understandably, that it would be fun to have a foal – but it would be wrong not to consider very carefully what is to happen to that foal when it is no longer a foal but a 'teenager', a 'young adult' and finally a fully mature horse. What will it be good at, and what sort of demand will there be for it? Or will it end up as just another candidate for the dog food factory because it has poor conformation, poor constitution, a terrible temper inherited from its parents or gained through bad, inexperienced handling by its breeders?

This sort of thing happens to *many* animals every year simply due to lack of thought in the original mating; perhaps an entirely unsuitable stallion was chosen for the breeder's mare, or the breeder did not know how to handle, feed or accommodate breeding stock – perhaps he or she had failed to envisage a definite slot for the foal, and therefore didn't choose a stallion which sired the right kind of stock, or (most importantly) one which did not suit the innocent mare involved.

Horses in our society nearly all have to work: entertaining us and giving pleasure to the human race is their lot in life today. Therefore, it makes sense to breed foals likely to grow up able to do this. Think about your mare first. What is she good at? Is she a comfortable, well-mannered and reliable hack? If that is all, you may think, 'she can't do anything'; but in fact the qualities she has are not that common. She must have a sensible temperament and be amenable to training to have become such a good hack in the first place, and this is an excellent start. To be a comfortable ride she must have reasonably good conformation, and this is essential for breeding working horses, and particularly high performance ones; and if she can jump a bit, this is a further plus point. Write down all her good qualities – and her bad ones because every horse has some – and consider them carefully.

You should also get an experienced consultant to assess her conformation and give you an idea of what type of animal she might breed when put to different sorts of stallion. This need not necessarily be a breeder, who may have fixed ideas of the sort of stock he or she breeds and may not like your mare, but it does need experience, a good eye and, ideally, a knowledge of your mare's personal qualities and temperament, so perhaps your instructor would be a good choice or someone experienced from your local riding club.

Land, buildings and other facilities
Accommodation for general riding horses has been discussed already in this book (p.39), and for breeding stock you simply have to look along the same lines; but make sure that everything is super-safe, as foals *do* get into trouble almost anywhere, and that your land is super-clean and well maintained (as foals are extremely susceptible to parasite infestation and disease), and that your stable is

not only clean but larger than a normal loose box – at least 14ft square for a mare and foal, as they live together for the first six months or so of the foal's life.

Your fencing in particular must be exemplary, good, thick hedging being the best, with post and rail fencing second. Wire fencing of any sort is not really suitable for breeding stock as it is not easily visible, and little feet can get through all but the finest chain link. The paddock must be well drained, with ample grass, and make sure there are no dangerous projections such as troughs set across a fence line rather than along it, no ponds, no farm machinery hidden in the grass and preferably with no access to vandals.

You will need the usual storage accommodation for feed and bedding.

It is a mistake to think of breeding a foal if you do not have sufficient land: to develop properly, foals *must* have a lot of freedom every day except in absolute emergency. Land which bakes hard in summer and is like a quagmire in winter is useless for breeding stock – even one foal needs to be able to stride out, run around and keep on the move in fresh air if it is to have any chance of maturing mentally and physically, into the sort of horse able to earn its living and find a buyer. The mare's health, too, depends on plenty of natural exercise on well drained, yielding land.

Company can be rather difficult for the one-mare owner. There is no doubt that foals, like children, do develop better physically and become better balanced mentally when they have others of their own age to play with. Lone foals often find it difficult to relate to peer groups once they are weaned, having missed out on experience of natural herd surroundings and manners. Obviously the dam will teach the foal a good deal, but it's not the same as learning to fight your own battles in the 'playground'. For this reason, it's often a good idea to keep your mare and foal at livery on an established stud, or at least with a friend who also breeds, even though you may be reluctant to do so.

General management

Probably the single most important aspect of management is good feeding, and we have an excellent part in this book written by nutritionist Gillian McCarthy. Basically breeding stock, particularly lactating mares (those giving milk) and those in the last three months of their pregnancy, and also youngstock in their first year, do need more protein than other horses.

A foal obtains all its needs from its dam for the first few weeks, but will quickly start experimenting with her hard feed even if it spits it out at first. Within a few weeks, the foal will require a feed of his own and then you must be careful that his dam does not chase him off it. On studs you will often find 'creep feeders', small enclosures which the foals can get into but the larger mares cannot; the foal's ration is fed in here so he can eat in peace. If you cannot run to a creep, you can either stay with mare and foal till they have eaten (excluding hay) or section off part of the loose box, if it is big enough, and put the foal in there during feed times. Provided he can still see the dam he will be alright.

Although you cannot wrap your stock in cotton wool, you should take no chances as regards safety. Choose a soft plastic feed container for the foal and remove it as soon as he has finished, so he cannot hurt himself on it, or chew and swallow bits of it afterwards. Hay is often fed loose on good studs, as this is the only way of ensuring there is no chance of injury on containers such as haynets, racks or hay holders. Foals can rear up and play in boxes and easily get their feet caught in the safest-looking equipment. If you feed your hay loose in two or three corners of the box you will ensure there is always a pile for dam and foal individually and that the foal cannot possibly come to any harm.

Breeding stock, particularly foals during their first winter when they will probably be weaned, do need adequate shelter. Foals should really come in at night during the first winter unless they are hardy, full-native stock, and even then I feel they should have a welcoming shelter shed to congregate in, with

BREEDING

What are you aiming for?
Land, buildings and other facilities
General management

HEREDITY

Like humans and other creatures, horses may take after their parents, or they may not. 'Throwbacks' to previous generations can mysteriously appear, but one thing is certain: genetically, your foal will have 50 per cent his dam's genes and 50 per cent his sire's – which are dominant depends on which parent's genes are the 'stronger' or more prepotent (dominant).

As mentioned, by carefully selecting, with expert help, a stallion to complement your mare, you can anticipate what your foal may be like. For example, if your mare has an over-long back and your proposed stallion an extremely short one, your foal will not have one somewhere in the middle – it is more likely to go to the extreme of one of its parents. You should, in this case, pick a stallion with an average, moderate-length back, good and strong, and particularly well made in the barrel region, and there is a better chance that this will be the prepotent factor which is passed on to your foal.

All other characteristics are inherited – coat colour, jumping ability, temperament and constitution, in fact just about every quality you can think of. To summarise, if you choose a stallion with particularly good points where your mare is weak, and also one who 'stamps his stock' (in other words, his stock all resemble him) – and if he has proved that he produces performance stock, or is a proved performer himself, then you stand a good chance of knowing more or less what your foal will be like: but you can never be certain, which is half the fun!

BREEDING

The natural approach
Choosing a stud
Mating

bedding and hay and preferably a nearby water source. It does not 'harden up' animals to expose them to harsh weather conditions, either a bitter winter, or a typically British wet and windy one, or an arid, searing hot summer. It merely stunts them. Extremes of climate and temperature and being forced to tolerate intolerable conditions are *not* what is wanted for breeding stock. So provide plenty of shelter all year round for all your stock if you want it to feel secure and content and to thrive, making the best use of the feed, care and attention you lavish on it.

Basically, the successful raising of horses requires good shelter, suitable accommodation, appropriate veterinary attention, exemplary feeding and worming, virtually limitless exercise and freedom, and also company – plus knowledgeable, common sense handling, and knowing instinctively when to call in expert help when you are in doubt.

The natural approach

Having said all this, readers may well be thinking about the feral herds of horses and ponies which live out all year round in wild conditions, and be wondering why extra feeding, buildings and so on are needed. The reason is that those feral horses are more or less adapted to their own environment – they are not going to be asked to develop to their maximum potential in order to please the human race; they will never have to gallop and jump across country, carry a human or pull a vehicle. Moreover the natural way does have disadvantages, in that many animals may die each year from exposure and disease, depending on the weather conditions prevailing. This is nature's way of selecting those most capable of thriving in a given environment.

We can use nature's basic aims and get horses to thrive in *our* environment by breeding the horses for the job, but we must recognise that the extra stresses we place on them are artificial stresses, and remember that they are basically the same horses as those running free on the plains where no extra demands are made on their resources. This makes two factors obvious:

1. We must conserve the domesticated animal's resources by feeding it well and providing adequate accommodation so that instead of using up energy to fight the weather, it can use it to develop to maximum potential.
2. We must allow for the horse's natural psyche, its physical functions and the mentality nature gave it by allowing maximum freedom and movement for breeding stock to develop and providing company for the development of mental maturity and stability. It is well known that animals who have learned natural herd manners are very much easier for humans to handle and train in later life.

It's true that many animals are not raised this way and seem fairly successful, but it would be interesting to know how much *more* successful they would have been if given a more appropriate upbringing.

Choosing a stud

You may be lucky, and find a stud which is very near to you, which both agrees with your idea of management and also has a suitable stallion. It is more likely however, that you will have to search the stud advertisements in the equestrian press, and ring or visit and speak to the owners to find out whether or not you are on the same wavelength. Take an experienced friend or adviser with you.

Particularly on studs breeding financially valuable animals such as Thoroughbreds (which are not all used for racing) and high performance horses, you may feel that the stallions at least are kept rather artificially and that the mating procedure, with twitched and hobbled mares, half a dozen attendants and no natural courtship allowed, is not what you would want for your mare. However, most stud owners play strongly on the side of safety and go to extreme lengths to avoid injury to their stallions. If you strongly disagree with this, simply

CHOOSING A STALLION

When choosing a stallion you have to consider not only the mare and her potential as a broodmare, but also the credentials of the stallion. These days, you should only send her to a stallion who is either a proven performer himself or who has a long and successful record of producing stock able to perform. Just because the horse is 'well bred' – a Thoroughbred, a 'good sort' and so on – does not mean he is going to sire something potentially useful, able to find a market slot and a good home. Horses today need 'papers' – registration papers with some suitable organisation (and this does not necessarily exclude 'mongrel' animals), either a breed society, or a performance register of some kind. There are many around, so check with your consultant as to what might be suitable.

If you choose a proven performance stallion of a particular breed you will be able to register your foal as a part-bred of that breed, which is a real step in the right direction.

If you work along these lines you should soon have a better idea of the sort of animal you want to breed, and will not be guilty of bringing into the world an innocent, sensitive animal with no future but the bullet.

142

keep trying till you find one which does things more naturally, particularly if your mare is an experienced broodmare and you would like her to run with her 'husband'. This is an involved subject, but generally many people feel (myself included) that they would rather have things done with more consideration for the animals' natural behaviour than is the norm, even if modern conditions cannot allow this altogether.

The main thing is that you feel happy with the place, that your adviser seems happy with the place, and that you feel you can talk to the people there. Most of all, you *must* sense that the animals feel content.

Get full details of the stallion/s you are interested in, particulars of the stud owners' requirements regarding your mare (swabs for disease, vaccinations, removal of shoes, season record which your vet will help you compile) and payment conditions for the stallion and your mare's keep.

Mating

Your mare will be mated according to the stage of her season, and this is something the stud staff and the stallion himself will decide; it is best done as calmly as possible. Some studs use a 'teaser' (a different, less valuable stallion) to test a mare by sniffing at her and nuzzling her to check her reactions. When she is ready she will spread her hind legs, lift her tail and 'wink' her vulva open and closed – but just as the poor mare has got interested in one chap, another completely different one is brought into to do the deed! Not really fair, but some stud owners feel it takes the strain off the 'star' by having his mistresses prepared in advance for him. Nonetheless, many top studs operate very successfully without it and – from my statistical research – seem to have higher conception rates; which is not really surprising, as the horses are less stressed.

The stallion will be brought in, should be allowed to say 'hello' to the mare and nuzzle her by way of enquiry, then will mount, perhaps at first from the side to avoid being kicked if he has any experience of fickle females, mate her, and be

In-hand mating, like this, is very artificial but often considered necessary because of the value of the animals. Possible upset to them can be minimised by quiet, sympathetic handlers.

allowed to dismount in his own time. In many studs warm water will be thrown over his penis (disinfectant being counter-productive) and he will be led away – and that's that. All that preparation for five minutes! She may be served two or three times during her season, depending on the stud's policy.

Of course, it is the forthcoming months and the foal which are important, and the real object of the exercise. The stud staff will carefully check the mare to see whether she comes into season again, which will obviously mean that the service was not successful and that she is not pregnant. They will liaise with you about further services and perhaps having to keep the mare till she does conceive. The process can be helped by hormone injections which are very successful, but do cost money. However, you should be consulted all along the way.

Back home
Broodmares can have normal exercise (though not *too* strenuous) once declared safely in foal (consult with your vet about checks necessary), but should not be ridden or driven from about eight months of pregnancy. They should be kept calm, turned out with quiet friends to get as much natural exercise as possible, but should not be stressed by weather, harrassment or undue poking and prodding.

When her time comes you would be well advised either to pay someone fully experienced to come and 'foal her down' as the term goes, or send her back to stud – which you may want to do anyway if you want her to be put in foal again. The stud staff are experts at foaling mares and, although you may feel you have to learn sometime, I think for your first it's better to be safe than sorry. If you really want to be present they may suggest you stay nearby, and will then ring you should she start to foal; it depends on the stud.

She will be mated again, if you have arranged this, either when she next comes into season (called the 'foal heat') or a season later; and will, you hope, be returned to your tender care safely in foal and with her foal at foot. But don't ever panic! Your vet will help all he or she can, and hopefully you will have other people on whom to call, plus a home library of books on breeding horses! Remember, foaling and growing up are perfectly natural processes in any species. The chances are that with correct preparation and knowing when to ask for help, everything will go swimmingly and you'll have nothing to worry about.

Weaning
In the wild, weaning takes place at eight to twelve months of age and is *very* gradual. In domestication, it is normal practice to wean at six months – and at this time, many mares and foals receive a severe mental and physical setback. It is still commonly felt that after six months the mare should be left to develop her resources for the new foal developing inside her, and that her previous foal should be able to manage. Gradually however, in some spheres at least, it is being realised that this is by no means always the best procedure. With modern feeds and sensible management, it is now known that a mare can very easily feed her existing *and* her future foal well enough up to eight months of age, and maybe even longer depending on the mare. The existing foal will be taking fewer and fewer sucks and so the demands on the mare are decreasing anyway, and it will be spending a good deal of time with its playmates rather than hanging on to Mum's apron strings.

Transportation
The same rules apply to breeding stock as for travelling other horses, with the exception that mare-and-foal units should be allowed to travel loose in the back of a horsebox together, not tied up or rugged in case they get caught up. The driver must drive *extremely* carefully and steadily. Only use a trailer (and a very large one at that) if you have absolutely no other choice, as they are neither as safe nor as stable as a proper box. Hire one if you have to, but do your utmost to use a horsebox for a mare and foal.

THE BOISTEROUS FOAL

There *are* cases – for example, when a boisterous foal is harrassing the mare, or the mare is bullying the foal or is not a good mother – when conventional or even early weaning are advisable, but normally weaning is carried out far too early. Many foals are scarred mentally for life at the severe trauma of losing their dams – it is a true bereavement for them, and for the mother – and if it is decided to wean the foal, at least do it gradually.

One method is to introduce a non-breeding 'nanny' mare or two into the field so the foal gets to know other adult mares. The foal and its dam can be separated for longer and longer periods each day over a period of weeks, the mare taken out, perhaps, leaving the foal with the nannies; this also helps the mare's milk to dry up gradually. Eventually complete separation can take place with the minimum of fuss. It is usually said that dams and foals should be out of sight and earshot of each other at this time, and although this may be difficult for 'small' owners to arrange, efforts should be made to send the mare to a friend for a few weeks; after this she can be reunited with her foal, if desired, with no ill effects.

Certainly *never* leave the foal alone. you must make arrangements for it to have another animal of preferably the same age, or at least a friendly mature horse or pony. Don't make the mistake of thinking a donkey or a cow are alright: they are not the same at all. Be fair to mare and foal if you want them to be content and thrive! After all, how would you like it?

Handling

I hope the photo sequence on handling foals will prove of help to readers – they say one picture is worth a thousand words! The main points to note are that it is never too early to start handling your foal properly. You don't have to give formal lessons, just do things right every time you have anything to do with the foal, and it will probably come to hand naturally.

It is dangerous and foolish to delay this process. Foals are very canny, and if they find they can boss or beat their humans, soon take advantage of it. If you never let it happen and never have a battle, you will have a co-operative, naturally mannered foal for your own pleasure, who is much more likely to find good owners in his long, future life.

Experts agree that the most trouble-free and effective way to teach a horse to accept handling is to begin when it is young – within reason the younger the better. Many foals come into the world and receive various degrees of handling from their owners or grooms from none at all, to speak of, to rough and detrimental treatment, neither of which augur well for the animal's future.

In this photo-feature, we have set out the correct and safe way to handle the young foal. The key points to remember are to be alert but relaxed and determined. If a foal becomes difficult, get on to it at once. Do not think: 'We'll give it a few days to settle down, then it will get used to us'. Those first few days of life are plenty for it to learn it can lead you a dance whenever you come near it. Even if it is your first foal, adopt a confident air. If there is any hesitation on your part the foal will spot it and play on it, so keep one step ahead of the foal – all its life.

Close boarding erected behind field mangers on a Thoroughbred stud to stop horses seeing and interfering with each other from neighbouring paddocks while feeding. These yearlings are wearing headcollars with short catching straps hanging from the bottom dees, obviously making it easier for handlers to quickly catch hold of them before clipping on the leadrope. Although it is common practice, segregating youngstock into peer groups like this is unnatural as they have no association with younger or older animals and do not learn their position in society. It is best to at least have one or more older horses with them, such as a barren mare or a gelding of mature years, to help teach them herd manners. This also indirectly induces more respect of humans.

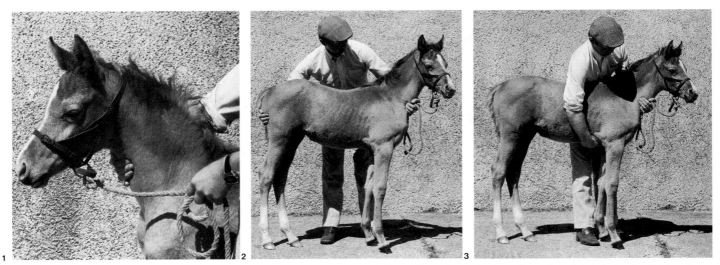

1 A well-fitting foal slip. The foal should be accustomed to wearing a slip from a few days old; indeed, many top Thoroughbred studs fit a slip from a few hours of age operating on the principle that the foal then never has to get used to it but regards it almost as part of itself, having always worn it.

The slip should fit snugly round the nose, like this, with enough room to let the foal move its jaws easily. The noseband should come an inch or couple of centimetres below the sharp facebones to avoid rubbing. If a slip is too large, a foal can get a hoof in it when scratching its head, and if too small it will be uncomfortable or cause soreness. When showing, the strap ends should be tucked in, as shown, but for general daily use they can be left loose for quick-release purposes in case of emergency. It is usually considered safer to remove the slip when the foal is not being handled, i.e. when stabled or turned out, although on big studs with visiting mares slips are sometimes left on for quickness when handling and catching and for ease of identification.

In this photo, the leadrope is shown tied firmly to the ring on the slip and can also have a knot tied in the loose end to prevent its being pulled through the hand. Some studs have a policy of leading mares and foals with double-length leadropes (often plaited binder twine) passed loose through the ring, both ends being held by the handler. This is so that should the animal break away and tread on the rope the rope will be pulled harmlessly through the ring, whereas with the other method the foal or mare could bring itself down by treading on a fixed rope.

2 This is the correct way to hold and guide a young foal, with one hand behind the quarters to urge it gently on and one round the neck/breast to prevent it charging off should it become over-enthusiastic about the 'aid' from behind. This is also the correct way to pick up a young foal for, for example, weighing. The handler picks up the youngster like this and steps on to the weighing machine, the total weight is read off, the weight of the handler deducted, and there you have the weight of the foal.

3 Never do it this way. This method can damage the heart and/or lungs through a foal's soft ribcage if one tries to pick it up like this. If trying to lead or guide this way, you have no control over the quarters.

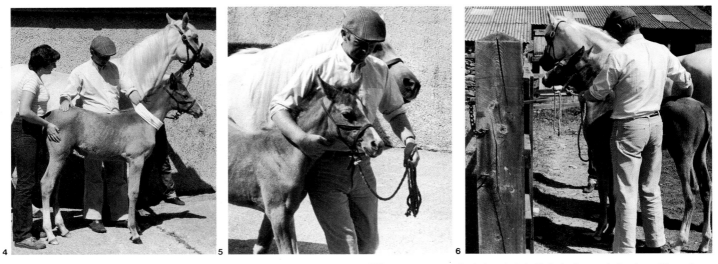

4 Early leading lessons begin with a stable rubber round the neck, as here. Preferably there should be three people, one leading the mare normally, one holding the rubber with a free hand to assist control of the quarters and a third behind the foal gently pushing and guiding. The third person is not essential but helpful for extra control as, come what may, the foal *must* learn at this early age that it has to go where it is directed. Once it learns it can say 'no' or get away, it will remember all its life that humans are not invincible and will be unreliable or even difficult to handle for the rest of its life. Many studs make a practice of leading the foal from different sides so that it does not become 'one-sided' and is used to being handled equally from both sides. It is also thought by many that this minimises the problems of 'sidedness' when a horse comes to be worked, so that it does not show undue stiffness or resistance on one side, usually the right.

The next stage would be to lead the foal without a third person behind, with one hand behind the quarters and the other holding the stable rubber instead of being round the breast or neck. A second person leads the mare. One hand is still behind the quarters to act as an auxiliary encouragement, and the left hand holds the leadrope but can quickly be transferred to the breast or under the neck for extra control.

5 More progress means the foal can be led with the leadrope and a hand round the neck like this, as a brake if needed. Loop the spare rope up in your hand but never, when leading any horse old or young, wrap it round your hand or wrist as you could be dragged this way should an animal take off. Eventually, one person can lead a mare and well-trained foal with both ropes in one hand, leaving the other free.

6 Going through a gate, the mare and her handler go first, the handler opening the gate if it is not already open. Mare and foal are led calmly through the gateway and turned into the fence like this, so that the foal feels itself still confined and does not charge off, knocking over attendants.

Release the foal first, then the mare, otherwise the foal might panic if it sees Mother going off into the paddock leaving it behind. Done this way, the foal will probably stay around the mare waiting for her.

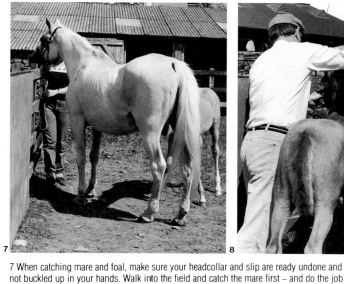
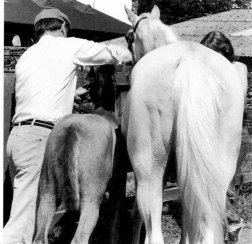
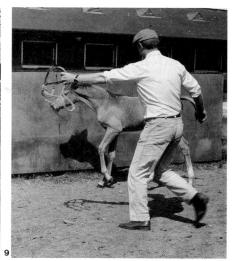

7 When catching mare and foal, make sure your headcollar and slip are ready undone and not buckled up in your hands. Walk into the field and catch the mare first – and do the job as if you *intend* to catch the mare and foal, not as though you are out for a Sunday afternoon stroll. Place the mare into a 'V'-shape with the fence, like this, with her handler blocking the gap under her neck, as shown.

8 One person, or two if available, can then calmly but firmly drive the foal into the 'V', as here. Stand with your body by the foal's quarters so it thinks there is someone behind it preventing it running backwards, and lean forward to fit the slip on its head.

9 If a foal is charging about, you can use your arms to make yourself 'as big as a bus' as far as the foal is concerned, and so very hard to avoid. It is no good catching up foals while you are half asleep. Keep your eye on the foal's eye to see what it is thinking of doing, so you can keep one step ahead of it, and use one or both arms to block it, as necessary. Sometimes, two people are not enough if the foal is very lively or reluctant to come in. If you 'misfire' twice, get a third person before trying again as a foal soon learns it can avoid being caught, and we want to keep such experiences to a minimum. Trying in vain to catch such a foal not only teaches it to play you up, it can also waste valuable hours on a stud.

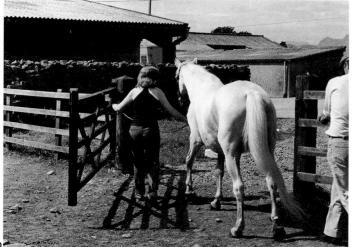

10 When leading out of a paddock, the mare leader again copes with the gate and the foal and its handler follow, like this, care being taken to guide the foal's quarters so that they do not bang against the gatepost. Do not take the foal in front of the mare until it is older and more experienced and has, perhaps, developed more independence as it inevitably will as it grows up. If you do, it might nap and play up, something we do not want it to know about.

11 Never pull on the leadrope like this as the foal might play up, rear or come over and hurt itself. If a foal starts playing up because of your mistake, bring the mare in front at once, and the foal will almost certainly quieten down.

VETERINARY CONSIDERATIONS

by Simon Wolfensohn, MA, Vet MB, MRCVS

REPRODUCTIVE PHYSIOLOGY

The mare

Reproduction in all mammals is controlled by hormones from three organs; the hypothalamus (part of the base of the brain); the pituitary gland (which lies below the base of the brain); and the ovaries. In the pregnant animal the system is also affected by feedback mechanisms from the foetus, placenta and uterus. The nervous system also affects reproduction, since it interprets the changes in daylight length and temperature which affect the activity of the mare's cycle.

The whole system is very finely balanced, and controlled to a large extent by hormones: the hormones are chemical 'messengers' which are carried around the body by the blood, and which work by the effect they have on their target organs, the ovary or uterus for example, and by producing feedback effects which switch their own production on or off. Many hormones can be used in the treatment of disorders, either in their naturally occurring form or as synthetic analogues.

Nerve cells in the hypothalamus produce a 'releasing factor' in response to light, smell, nutritional factors, temperature and, in many animals, sexual stimuli such as mating displays, chemical signals (pheromones) or mating behaviour. The releasing factor stimulates the pituitary gland to produce the hormones FSH (follicle stimulating hormone) and LH (luteinising hormone); these act on the ovary to make it produce eggs (ova) which are then shed into the uterus via the fallopian tubes. The tissue (follicles) from which the eggs were produced then matures to form a structure called the corpus luteum. While maturing, the follicles produce the hormone oestrogen, and this acts by feedback to the pituitary gland to shut off hormone production. The corpus luteum also produces feedback effects by the production of the hormone progesterone. Oestrogen acts on the uterus to prepare it for conception, it causes the mare to display oestrus behaviour (season) by its effects on the nervous system, and feedback from the ovary is important in the timing of ovulation. Progesterone also helps to prepare the uterus for the implantation of the foetus, helps to maintain the right conditions in the uterus for pregnancy, and inhibits further oestrus cycles by its action on the hypothalamus. Other factors are also involved, but these are still not well understood.

In the non-pregnant mare another type of messenger, prostaglandin, is produced in the uterus and this causes the corpus luteum to break down and allows further seasons to develop. The presence of a foal prevents this and allows pregnancy to continue.

Both the follicles and the corpus luteum have the capacity to form cysts which may interfere with these feedback mechanisms. Feedback failures from other components of the system may also result in the mare failing to come into season, or to ovulate, conceive or maintain pregnancy. Such failures may be due to many causes; these include external factors such as daylight length or feeding, or physical factors such as abnormalities of the reproductive organs or foal. Infections of the reproductive tract may also interfere with fertility, when the organs may be directly affected, or the hormonal mechanisms.

The stallion

Successful mating requires the desire and ability to mate, and the production of semen containing adequate numbers of normal spermatozoa. The male animal is affected by temperature, daylight, nutrition and environmental factors (including the presence of a mare in season); his performance may be affected by disease, and he is also under the control of hormones: FSH controls the production of sperm, and LH controls the production of the male hormone, testosterone. Male characteristics, sexual desire (libido) and the production of semen to carry the sperm are controlled by testosterone. There is a feedback of testosterone to the pituitary gland and this regulates its production.

THE REPRODUCTIVE CYCLE

The first season is usually seen when a filly reaches her second summer, and she would not normally produce a foal before the age of three. The mature mare has a period of sexual inactivity through the winter, then her seasons begin again in the spring and continue through to the autumn. A few mares will show oestrus cycles all year, especially if given extra food and warmth through the winter months. At the start of the breeding season the mare's heats are rather variable in length and character, and may not be accompanied by ovulation.

The reproductive cycle is divided into five phases:
Pro-oestrus: This phase precedes the visible signs of oestrus. Activity of the reproductive organs increases and the follicles in the ovary develop.

Oestrus: During oestrus the mare will accept the stallion. A mucus discharge is produced by the vagina and the vulva becomes enlarged and mobile.

Metoestrus: The corpus luteum forms in the ovary and the uterus becomes active to prepare for implantation of the fertilised egg.

Dioestrus: this is the resting period between successive heats.

Anoestrus: anoestrus is the inactive phase between breeding seasons.

The mare's oestrus phase lasts for six days and the whole cycle is 20-23 days with ovulation occurring on the last day of heat or the day before. After foaling, heat occurs on the eighth to tenth day. This 'foal' heat may be short, only two to four days, and it is common practice to cover the mare again on the ninth day after foaling. The foal heat is sometimes more obvious because of the foal's diarrhoea than because of the mare's oestrus behaviour.

For mating it may be necessary to keep the foal out of earshot, or in other cases to keep the foal restrained close by when the mare is teased or served – some mares with foals at foot are naturally anxious about the foal, and oestrus behaviour will be inhibited.

When she comes on heat the mare becomes restless and irritable. She takes up the stance for urination frequently, and when she passes urine she turns the lips of the vulva outwards and exposes the clitoris. When a stallion is present these signs are more exaggerated and she may raise her tail to one side and lean her body. The vulva will be swollen and there will be a discharge of mucus.

Mares that are not in season will object violently to the advances of a stallion, and therefore a barrier should always be put between them while trying to see whether the mare is willing to accept him. On studs a 'teaser' is often used to check whether a mare is in season; ideally the mare should be teased every two days to see whether she is ready for the stallion. In some cases a mare which reacts favourably to the teaser will still object to the stallion (when there is no barrier) and for this reason the mare is often fitted with felt boots or hobbles to prevent injuries from kicking; a twitch may sometimes be necessary. When a mare is in season a stallion will usually show typical behaviour called 'Flehmen's posture' named after the man who first described it; this means raising the head, flaring the nostrils and raising the upper lip.

The effect of daylight

Daylight is very important in regulating sexual activity in several species. To prepare a mare for early breeding she should be exposed to increased light from January 1st, and artificial light should be supplied, by a 200 watt bulb or equivalent in each box, in order to provide sixteen hours of light and eight hours darkness. It is possible to buy fluorescent tubes with a light emission spectrum almost exactly equivalent to normal daylight, and these may be better than incandescent bulbs. It is very important to continue this light treatment at home as well as when the mare goes to stud, and it should continue for fifty days,

which will therefore end on the tenth day of the covering season (starting on February 15th for British Thoroughbreds). Other schedules of treatment may be advised by your vet in particular circumstances, and light changes may also be combined with hormonal therapy. For successful induction of early oestrus it is also essential to provide good nutrition and extra warmth.

STUD HYGIENE

In order to avoid the spread of infection from one mare to another, it is a good idea for anyone who handles a mare, especially around the vulva or anal area, to wear disposable plastic gloves. Each loose box should be provided with a disinfectant tray outside the door, and boots should be dipped before entering the box and on leaving it. Ideally clean overalls should be used for each mare and there should be facilities of some sort for washing hands before moving on to another horse. All utensils, buckets and other equipment should be sterilised; if that is not possible, they should at least be cleaned thoroughly and reserved for use with one horse only. With an infected mare these precautions are essential.

The number of people handling a particular mare should be kept as low as possible, and visitors should not be allowed to roam about from one horse to another. On large studs a central veterinary facility is desirable, with a loose box and preferably stocks for veterinary use only, together with washing facilities for hands, boots, overalls and utensils. A steriliser for gynaecological equipment is useful, so that contaminated instruments need not be taken out of the area. Some bacteria can survive in normal concentrations of antiseptic or disinfectant, and because higher concentrations may injure the tissues of the mare or stallion, it is essential to use fresh solutions on each animal; the vet will therefore need a fresh bucket to use for disinfectant at each examination. Some vets, however, may prefer to use a spray of disinfectant, as this can be carried from box to box and avoids the risk of using contaminated buckets.

For mating, the vulva and surrounding area of the mare are normally washed with antiseptic, and some studs wash the stallion's penis also. It is very important that all traces of disinfectant or antiseptic are washed off, to avoid any contraceptive effect. Antiseptics may be harmful, too, because they can alter the normal bacterial population of the horse's skin – for these reasons, some vets prefer to use clean warm water without antiseptics. Immediately after mating it is usual to wash the stallion's penis thoroughly to remove secretions.

There are many disinfectants suitable for washing off boots and other items; ideally the choice should be made in consultation with the vet and in the light of any particular infection problems. Probably the most common chemicals for use on animals are cetrimide, chlorhexidine, or povidone-iodine, used as recommended by the manufacturers.

VETERINARY EXAMINATION OF THE MARE BEFORE BREEDING

Veterinary examinations of mares have several purposes:
  i. To find out the time of ovulation in order to maximise the chances of successful mating;
  ii. for early recognition of problems associated with cysts in the ovaries;
  iii. to find out the sexual state of a mare, *eg* anoestrus, oestrus, pregnant, etc;
  iv. to diagnose infection or other diseases.

The methods available to help with these questions are rectal palpation, vaginal examination, cytology, biopsy, hormonal investigations through blood sampling, and ultrasound scanning.

Rectal palpation

In this method the veterinary surgeon carefully inserts a gloved hand through the horse's anus, and uses his fingers to feel the structures of the reproductive organs through the wall of the rectum. This can be carried out around the stable door, or by using stocks – a solidly made and fixed wooden framework in which the horse stands, with a back plate to protect the vet from being kicked. In most places the vet will have to use the first method, as stocks are not often available. The mare is restrained with her hindquarters at the doorway, one person holding her head and another holding her tail to one side and steadying her hindquarters. The vet stands with his body behind the doorpost as far as possible, and reaches round it to carry out the examination. Most mares will stand quite quietly, especially once the vet has inserted his hand; however, some may be very difficult to examine and sedation may be necessary.

This form of examination allows the vet to feel the uterus and cervix as well as the ovaries; he can tell whether the mare is pregnant, and can also examine each ovary for the presence of developing follicles or a corpus luteum, or for follicular or luteal cysts. As far as ovulation is concerned, the advantage of this method is that the veterinary surgeon can manually rupture cysts if it is appropriate to do so.

The main disadvantages of this method are that it relies on the experience and skill of the vet, and it also presents some danger to the horse. The rectum of the horse is a fairly flimsy structure and, especially in tense or nervous horses, it is possible to cause haemorrhage from the lining of the bowel – in extreme cases the bowel wall might actually be ruptured. In nervous or uncooperative animals it may be impossible to make a diagnosis at all by this procedure.

Vaginal Examination

This is usually done *before* carrying out a rectal examination, which will lead to contamination of the area with faeces; the vulva and surrounding area are usually washed and dried before beginning. The vet uses a vaginal speculum, a clear plastic tube with a light and handle, through which he can see the cervix and vaginal lining of the mare. This sort of examination would also enable him to find a number of other conditions, such as pneumovagina or various infections, as well as finding out about the stage of the oestrus cycle. The results of visual inspection of the cervix are usually only interpreted in the light of findings from rectal palpation and possibly blood samples.

Cytology/Biopsy

Further information about the reproductive cycle can be obtained by taking samples of the cells found in the lining of the vagina, and looking at them under the microscope. The cell types found are characteristic of the stage of the oestrus cycle.

If the vet is investigating for infertility, he may also take an endometrial biopsy – a sample of the tissue in the lining of the uterus. It is obtained by passing a pair of fine forceps through the cervix, via the vagina, and nipping out a piece of the lining. The forceps are guided by the vet's other hand, which can locate the uterus through the rectal wall. This is often the only method which allows a positive diagnosis of some conditions affecting the uterus.

Blood samples

Hormone levels in the blood – for example, progesterone – may be helpful in assessing the stage of the oestrus cycle, or in the diagnosis of pregnancy. As a research tool, such samples are essential.

Ultrasound scanning

Modern ultrasound scanners can give a great deal of information about the ovaries, and also the state of pregnancy. A probe is inserted into the mare's rectum and directed towards the structure under examination; put simply, a beam of ultrasonic sound which has a frequency far above the audible range, is directed at the tissue. Some of this sound energy is bounced back from the tissue and is picked up by the the probe: the reflectivity or 'echogenicity' of the tissue is characteristic, and is analysed with the aid of a computer. The results can then be displayed as a picture on a screen, or printed out using a video printer.

Ovarian cysts can be recognised by this method, ovulation can be pinpointed, and other abnormalities such as ovarian tumours can also be found.

STALLION MANAGEMENT AND FERTILITY TESTING

How fertile a stallion proves to be depends on a number of factors: its management in terms of feeding and general care; psychological factors; over-use; physical problems in mating; and the management of mating itself – all these may influence the outcome. On the other hand, attempts to mate with subfertile mares may sometimes look (statistically) as though the stallion is performing poorly.

Genetic factors may lead to poor semen production or damaged sperm cells, or there may be abnormalities of the chromosomes – all these can cause infertility. Horses with chromosome abnormalities usually have visibly abnormal genital organs and are therefore not usually expected to breed; nor are horses with other inherited abnormalities such as umbilical hernias or retained testes.

There is no evidence that hormonal problems are responsible for infertility in the stallion, and treatment with hormones does not seem to produce any long-term improvement. In some cases an injection of LH before service has been found to increase the stallion's sexual desire.

Infections, injuries and tumours may all be found affecting the penis, testicles and accessory glands of the stallion.

Management

A seriously overweight or underweight stallion will have poor fertility. Good housing, warmth and general care are essential to his well-being; and although less crucial than for mares, daylength is of some importance as he, too, is expected to be sexually active earlier than is natural. On a stud, the stallion will only be allowed to mate once or twice with a particular mare (whereas in the wild he would probably mate several times) but he will be expected to cover more mares; therefore the timing of mating is very important to avoid over-using him. The covering season is quite short, and some stallions are simply expected to mate more times than is reasonable. A 'book' of forty mares is normal for one season for each stallion, although many are expected to cover a lot more.

Nowadays, horses are trained and kept in a disciplined environment where they are expected to obey commands; however, a stallion entering on a breeding career may easily be confused by rough handling, or handling which leaves him unsure of what he is supposed to do. It is important that he has a good relationship with the stallion man, and also a gentle introduction to the process of mating – a mature mare should be used, well in oestrus. Failure to achieve successful mating should be ignored initially, and the stallion should be allowed to rest for several hours before trying again.

A stallion may be attributed with infertility when in fact it is the stud manager who fails to recognise ejaculation. One of the studmen should confirm that ejaculation is taking place by resting a hand lightly on the underside of the penis – he will feel the pulsation as the stallion ejaculates.

Psychology

In addition to the factors already mentioned, it sometimes happens that a stallion will take a dislike to a particular mare and show little interest; in some cases lungeing him for a while seems to improve his desire. Other stallions may take an excessively long time to mate, and this may limit the number of mares which they can cover.

Other problems associated with behaviour include failure to maintain erection of the penis, difficulty in entering the mare or failure to ejaculate, or dismounting too soon. In some cases the situation may be improved by allowing the stallion and mare to run free, or by resting the stallion.

Evaluation of semen quality

Normal semen varies considerably between stallions; it also varies depending on the time of year, how much the stallion is used, and the methods of collection. Factors such as the general health of the stallion and the number and quality of mares covered should also be taken into account before considering the stallion to be infertile.

Measurement of the testicles allows a rough calculation of the expected output of sperm (each gramme of normal testicular tissue produces 20 million sperm daily), and roughly 80% of this can be collected by artificial methods. If the actual output is considerably less than the expected production, an abnormality of the testicles may be suspected.

The only satisfactory way of obtaining a sample of semen is to use an artificial vagina. This is a tube surrounded by a water jacket to warm it, and the stallion's penis is guided into it as he attempts to mount a mare; the artificial vagina ends in a collection vessel of some kind.

Information about semen quality is based on the appearance of the sample, its volume and its acidity, and on the presence of other cells in the sample (such as red blood cells); also on bacteriological examination as well as the characteristics of the sperm.

The individual sperm are made up of a head, which has a cap called the acrosome, and a tail and they look rather like microscopic tadpoles. They move by propelling themselves with the tail, and the acrosome is important in penetrating the wall of the egg in order to allow fertilisation. Significant characteristics to look for when examining the sample are the concentration of sperm; the ratio of live to dead sperm (assessed using chemical methods on slides looked at under a microscope); the motility of sperm, and therefore their ability to move through the reproductive tract and reach the egg for fertilisation; and the presence of abnormal sperm, such as ones with missing or shortened tails or defective acrosomes.

Many of these assessments are rather subjective and depend to some extent on the experience of the person carrying out the examination. Nor is prediction of fertility a very accurate procedure – for example, it is sometimes found that stallions with apparently poor semen quality are adequately fertile as long as they are well managed.

DISEASES AND DISORDERS OF BREEDING STOCK

Chromosomal abnormalities: A very small number of mares have abnormalities of their chromosomes which make them infertile; both their ovaries and the lining of the uterus are abnormal. It is possible that chromosome abnormalities are responsible for some otherwise unexplained abortions.

Cryptorchidism: A horse with one or both testicles retained within the abdomen, instead of descended into the scrotum, is described as 'cryptorchid'; the colloquial term for such a horse is a 'rig'. These horses often show exaggerated male sexual behaviour, probably due to increased hormone secretion from the retained testicle. The condition is inherited and rigs should not be used for breeding, so it is usually best to castrate them.

Endometritis: Various types of infection may affect the lining of the uterus, including Contagious Equine Metritis (CEM) and other bacterial infections. Many of the bacteria involved are normal inhabitants of the skin or environment, and enter the body either at service or when the mare foals. Some cases may be severe and rapidly developing, while others may be much longer term problems. Infections such as these are an important cause of infertility; they may be diagnosed by isolation of the bacteria responsible, or by biopsy of the uterine lining (ie removal of a small piece of tissue using specialised forceps). Treatment consists of antibiotics and irrigation of the uterus.

Equine coital exanthema: is caused by equine herpes virus 3, and is a venereal infection of mares and stallions which produces ulceration of the genitalia. The ulcers will usually heal in a few weeks, but treatment to prevent secondary bacterial infection may be needed.

Infertility: There are many reasons why a mare or stallion may be infertile: these may be an infection involving the reproductive organs, generalised infection, or a malformation of the genital tract; it may be due to an abnormal oestrus cycle, or tumours, even poor management of the breeding stock (eg failure to serve at the appropriate time); there may be physical or psychological problems which interfere with mating; and fertility can be affected by nutritional deficiencies. Early death of the embryo may also appear as infertility. (See also specific conditions: chromosomal abnormalities, endometritis, ovarian cysts and tumours, physical abnormalities, pneumovagina, psychological problems, stallion management).

Ovarian cysts: Cysts may develop in the ovarian follicles or in the corpus luteum, but how critical they are depends on the stage of the oestrus cycle and the structure affected. The vet's diagnosis is usually based on feeling the cysts through the wall of the rectum, or by ultrasound examination; treatment is usually by manual rupture of the cysts or hormonal treatment.

Ovarian tumours: Tumours of the ovary are fairly rare. They give rise to various sorts of aberrant behaviour including nymphomania or stallion-like behaviour patterns; in some cases the mare will remain in anoestrus. The most common type of tumour is benign, and surgery may be possible to remove the affected ovary. After surgery the normal ovary may function properly and conception may be possible.

Physical abnormalities: Physical problems which may result in the mare or stallion failing to mate successfully are quite numerous. They include simple disparity in size; physical deformities; painful conditions affecting the horse's ability to bear weight; lameness; or abnormalities of the reproductive tract such as deformities of the vulva or a persistent hymen. Skilled handling and veterinary examination before mating may help to overcome some of these problems.

Pneumovagina/Caslick's operation: A common cause of infertility is a misshapen vagina and vulva which allows air to enter. Sometimes the air can be heard entering the tract, and the common name for this condition is 'wind-sucking'. This leads to bacterial infection of the uterus and affects fertility. The problem can be corrected in many cases by Caslick's operation: the lips of the vulva are trimmed and sutured together so that only a small opening is left for the passage of urine.

Mares which have had this operation may need to have the vulva reopened for mating. Under local anaesthetic the vet can cut the vulva along the line of the operation scar. In preparation for foaling this can be carried out in the same way or it can be done without local anaesthetic by an experienced stud groom during the process of birth.

Psychological problems: A horse's ability to mate successfully may be profoundly affected by its temperament. For example, some mares are so nervous of the stallion that they will not stand successfully even when correctly in oestrus. Others may fail to raise their foal successfully because they turn out to be impossibly bad mothers. Stallions can often be extremely temperamental, and it is widely recognised that they should generally be handled by one person only, and that their relationship with that person is very important for successful breeding.

Tumours of the penis: Tumours of the penis are rare, although several types have been recorded. Tumours of the testicles are also rare.

PREGNANCY DIAGNOSIS

Several methods are available:
i. Rectal palpation; ii. Blood tests; iii. Urine tests; iv. Ultrasound.

Rectal palpation

The same technique is used as described on p.00 and the same restrictions apply. The advantage of the method is that in some cases an experienced vet can make a diagnosis from about the 18th day of pregnancy, especially if an

It can be very difficult to tell when a mare is pregnant just by looking at her. This mare was eight months pregnant when this photograph was taken.

examination has also been carried out before mating. In a small number of mares, however, the lower limit for diagnosis may be around 50 days. This method does not give any information about the health or otherwise of the foetus unless repeated examinations are made. Vaginal examination of the cervix may also give useful evidence to confirm pregnancy.

Owners sometimes believe that rectal palpation is dangerous to the pregnancy, but there is no evidence that careful rectal examination is a cause of abortion in healthy individuals. Very occasionally a few mares abort after this procedure, but it seems likely that they do so for other reasons.

Blood tests

Blood samples are useful from about the 40th day. They rely on the detection of a substance known as Pregnant Mare's Serum Gonadotrophin (PMSG), and modern tests use immunological methods (not live animals) which are very accurate. They only require very small quantities of blood and give rapid results.

An assessment of blood progesterone levels is only helpful if these are very low, which will confirm non-pregnancy – higher levels do not distinguish between the dioestrus stage of the cycle and pregnancy. These tests employ methods similar to those in use for PMSG.

Urine tests

Urine tests are only useful from about the 150th day. Rectal palpation and ultrasound are quicker and easier methods, and urine tests are seldom used these days.

Ultrasound

Ultasound is useful in diagnosing pregnancy from about the 80th day. As machinery becomes more sophisticated, diagnosis can be made progressively earlier – some machines can detect the first signs of a swelling formed by the foetus as early as 15 days. The technique depends on being able to differentiate the swelling of a developing foetus from cysts or other swellings of the uterus. An early diagnosis can be confirmed once the foetal membranes can be seen, or when the foetal circulation becomes visible.

FOALING

We do not understand fully the factors which initiate the birth of a foal, but we do know that the mechanism is not the same as that of the cow or sheep. The adrenal glands of the foal enlarge significantly near the time of birth, and researchers therefore think that steroids from these glands are important; however, the other hormonal mechanisms are not known. Signs of imminent birth are not reliable. The appearance of a waxy secretion at the teats is often a sign that foaling is about to happen, but some mares show this 'waxing up' two or three weeks before foaling. Even mares which start to produce milk may not foal for some time.

Labour

In the wild, or under natural conditions, a mare would detach herself from the herd and give birth during the night. When kept housed she will become restless, go off her food, and pace around her box, holding the tail to one side; her vulva will have become swollen and lengthened. She will produce dung frequently, and it will get softer; some mares may seem rather colicky, looking round at the flank and kicking at the abdomen, and they may sweat profusely.

As labour begins, the foal will turn from a position in which its abdomen faces the mare's spine, through 180 degrees until it is lying with its back nearest the mare's spine, and the head and feet nearest her tail. As the foal's feet are pushed through the mare's cervix by the contractions of the uterus, the allantoic sac will rupture and a gush of fluid will escape (the allantoic fluid is the foetal urine which has accumulated before birth). Sometimes the allantoic fluid will contain solid structures known as 'hippomanes' – these are flat discs anything up to about 4cm thick and 20cm in diameter; they are also

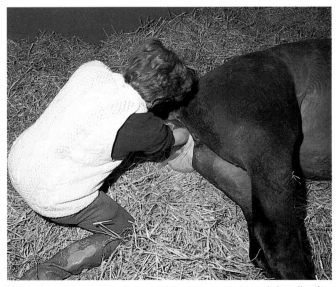

The amnion sometimes breaks during birth: if not the attendants can do it to allow the foal to start breathing.

known as 'foal's-bread' or 'foal's-tongues'. As the foal's head enters the mother's pelvis, reflex action causes the mare to start straining, using her abdominal muscles strongly to push the foal out. Most mares will lie down on their side as this second stage of labour begins, and the foal is usually delivered rapidly, still covered in the inner layer of membranes, the amniotic sac. The feet should appear first at the vulva, one slightly in front of the other, followed by the nose. The head will follow as straining continues, and the shoulders and chest should appear very soon afterwards.

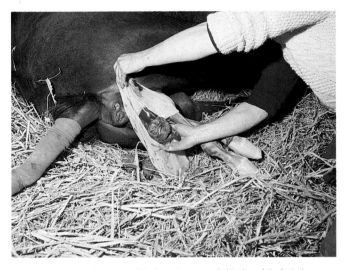

If a mare is having difficulty, possibly due to incorrect positioning of the foal, the veterinary surgeon can help her.

Mares sometimes pause once the chest of the foal has been delivered, and at this stage the amnion can be broken. A normal foal will breathe at this point, and fluids from the stomach will drain from the mouth and nostrils. The rest of the body is then delivered, although the hind feet usually remain within the vulva for a few minutes while blood from the placenta drains back to the foal through the umbilical cord which is still unbroken. It is thought that this is also a mechanism to plug the mare's vulva, thus preventing a sudden inrush of air which would contain bacteria and debris. It is important not to break the cord at this point manually as excessive bleeding may be caused; leave the cord to break naturally when the foal withdraws its feet or the mare gets up. The mare will then lick and nuzzle the foal and encourage it to its feet; some maiden mares, however, may be totally disinterested, or even aggressive to the foal.

After a time, which may vary from half-an-hour up to a maximum of ten hours, the foetal membranes should be expelled, possibly with some further straining. The membranes should be delivered inside out, and should be the shape of a pair of trousers, one leg of which will be larger than the other: the larger side lined the horn of the uterus in which the foal lay, while the smaller contained only membranes.

The actual delivery of the foal is rapid and violent, and takes only about twenty minutes, although the whole process may take several hours from beginning to end. Thoroughbred mares are kept under constant watch until foaling has been completed, as the birth may easily be missed if the attendant has only an hour off.

Once abdominal contractions have begun any delay in delivery is serious, and the vet should be called at once. Any abnormal presentation such as the head appearing before the feet, or the foal apparently being upside down, is also potentially serious and again you should call your vet.

POST-NATAL CARE OF THE MARE AND FOAL

'Stitched mares'

Mares which have had a Caslick's operation previously to correct wind-sucking will need to have the vulva reopened by cutting before delivery. This is usually done by the stud groom just before the forefeet make their appearance. At this stage the vulva is fairly insensitive, and a straight cut with scissors will not need local anaesthesia. In some cases this will be done by the vet a few days before foaling, when local anaethetic is used. After foaling the vulva will need to be restricted by the vet.

Looking after the foal

Most foals will escape from the membrane of the amnion in the course of delivery, but a few may need help, especially if weak or premature; if necessary, you can easily tear the amnion with your fingers.

A newly-foaled mare gets to know her foal by smell and taste at first. This bonding process is vital.

Immediately after delivery some stud grooms like to dry the foal using straw or towels. In most cases this is unnecessary, and it may also inhibit the mother's reaction to the foal by masking its smell, so a little discretion is needed. On very cold days it may be useful to prevent chilling.

At most foalings there is no need to interfere with the umbilical cord; it will usually break naturally and bleeding will stop without help. If bleeding from the cord continues, it is often enough to pinch the end with your fingers for a few minutes, though in some cases it may be necessary to tie it off with a ligature of clean tape.

Very occasionally the cord may not rupture by itself when the mare gets to her feet. If necessary it can be torn quite easily by hand, though take care not to pull excessively on either end.

A normal foal will stand within an hour and a half of birth, and should suck within three and a half hours; and once on its feet and sucking, the foal should be increasingly well coordinated. However, if these timings are not met your vet should be consulted, as any delay in dealing with a weak foal may be serious. The foal should be watched to see that it is sucking properly, and the mare should be checked to see that she has a good supply of milk. You will find that if the foal has been sucking within the previous half hour or so the udder will feel empty and slack but the teats will be wet. In contrast, it is easy to tell if the udder is full – it feels taut, and milk can be readily expressed by squeezing the teats.

It is better not to handle foals before they take their first suck, but if this is really necessary, any handling should be positive but as gentle and quiet as possible. Foals should be picked up by putting one arm around the front of the shoulders and the other around the back legs; they should not be lifted under the belly. If it is necessary to restrain a foal with convulsions you should sit on the ground with the foal lying flat and its neck across your lap. Restrain the foal's head against your body holding the upper leg at the same time. The same position can be used to feed a foal by hand.

If a foal does not breathe promptly after delivery it is possible to give artificial respiration by blowing into one nostril while holding the other shut; and if oxygen is available, it can be given by tube into one nostril. A breathing rate of about six breaths per minute should be used. If artificial respiration is needed you should call your vet, as further attention is almost certainly going to be needed.

In the period after foaling the mare should be observed regularly for signs of haemorrhage, abnormal discharges, weakness, lack of appetite, failure to produce milk, or for any other signs of illness (mentioned under specific disease headings). She should also be carefully watched to make sure that she shows good maternal behaviour towards the foal and does not attempt to injure it or refuse to feed it.

DISEASES OF THE NEWLY FOALED MARE AND FOAL

Angular limb deformity: Bone growth is controlled by areas at the ends of the long bones called growth plates. Damage to these areas may cause angulation of the limb which may be noticeable at birth or may develop later. Most minor cases will improve without treatment, but some cases need surgery.

Atelectasis: In some foals the lungs fail to expand normally at birth. Some may respond to treatment, but severe cases may die soon after delivery.

Contracted flexor tendons: Some newborn foals show a contraction of the flexor tendons, usually of the forelegs and less commonly of the hind legs. Some will improve with age as the tendons stretch, but others may need plaster casts or surgery and these may never be completely sound.

Haemolytic disease of the newborn foal: An immune reaction between the foal's red blood cells and antibodies from the colostrum can result in destruction of the red cells, and consequently anaemia. It is more common in Thoroughbreds, and it occurs because the foal inherits a blood type from the stallion which the mare does not share; she then produces antibodies to it as a result of foetal cells leaking into her circulation across the placenta. Signs usually develop at about 12–36 hours after birth: the foal becomes weak, pale and unwilling to suck; there may be visible jaundice, and some foals will die very quickly.

This condition can be treated by transfusions of specially prepared red blood cells, or of compatible blood. In order to avoid the condition the mating should not be repeated, or the foal should be prevented from taking colostrum from the mother and given some from another mare. A blood test on the mare in the last month of pregnancy may be helpful.

Haemorrhage: During foaling haemorrhage may occur from the uterus, either internally or externally. Minor haemorrhages may stop without treatment, but others may be due to tearing of the uterus or damage to larger blood vessels. Haemorrhage may also occur in the ligaments of the uterus, and this – or tears of the uterus – may produce signs of colicky pain with weakness and shock. Although the bleeding may be internal with no visible sign of blood, the loss of blood may in fact be major, and some mares may die as a result of these injuries. Veterinary attention is essential. Some blood is always seen at a normal foaling, of course, but it is important to realise that this is mainly blood loss from the placenta – under normal circumstances the mare does not suffer any significant bleeding. Therefore any visible bleeding should stop quite soon after delivery.

Lactation tetany: Mares may suffer from lack of calcium, especially ten days or so after foaling or a few days after weaning: signs include incoordination, sweating and muscle tremors. The condition was more common in draught horses and is quite rare these days. Treatment involves calcium solution given intravenously, and is usually effective.

Mastitis: Bacterial infection of the udder is possible at any stage of lactation. The udder becomes hot, swollen and painful and the mare may be unwell and off her food. Treatment with antibiotics is often given both by injections and into the teats.

In more severe cases it may be necessary to prevent the foal from sucking and to bottle-feed it for a few days.

Neonatal maladjustment syndrome: Foals suffering from this condition are sometimes referred to as 'barkers', 'wanderers' or 'dummies'; the causes are not completely understood, but circulatory problems and lack of oxygen to the brain are involved. The condition is more common in Thoroughbreds. Signs usually develop within twenty-four hours of birth and include convulsions, aimless movements, inability to suck, periods of coma, exaggerated reaction to noise, unequal pupils and abnormal 'barking' sounds. Some foals will recover if fed with a stomach tube, nursed and given treatment for specific symptoms.

Pervious urachus: The bladder of the foetal foal communicates with the allantoic sac via the umbilicus (navel). At birth this tract does not always seal itself properly, and urine may be seen dripping from the navel for a few days after birth. Most cases resolve spontaneously as the foal grows, but a few may need surgery.

Prolapse: of the uterus is more common at the first foaling than in older mares. The uterus turns inside out and protrudes through the vulva, usually after foaling but on rare occasions vaginal prolapse can occur before delivery. The prolapsed uterus is a large mass, possibly reaching almost down to the hocks, and it will have a bloody, rough appearance. The vet should be called immediately and the mare should be kept standing; if possible the uterus should be supported on a clean sheet and bathed with saline solution.

Retained membranes: The foetal membranes and placenta are normally passed immediately after foaling, and should certainly have been expelled within ten hours after delivery. If this has not happened you should call your vet. Manual removal of the membranes and treatment with antibiotics are usually necessary, although it may not be possible to complete the process for two or three days. If retained membranes are not removed, an infection of the uterus will probably follow, sometimes complicated by septic shock and laminitis.

Retained meconium: While still in the uterus the foal produces faeces which accumulate in the rectum and colon. This material is called the meconium, and after birth it has to be passed; but because there is a lot of it and it has a rather hard, rubbery consistency the foal often finds this difficult. Signs of abdominal pain may develop, with the foal straining to pass the meconium, and treatment by enemas, laxatives or manual removal of the faeces may be necessary.

Septicaemia/joint ill: (See also 'Sleepers'.) Any generalised blood-borne infection is called septicaemia, and a specific case of this is 'joint-ill', seen in young foals in the first few weeks after birth. Several different bacteria may be responsible which all become established in the joints; there may be generalised illness, and the foal will be lame with hot, swollen and painful joints. Infection often enters the body via the navel. Damage to the joints may remain even after the infection has been treated successfully with antibiotics.

Good hygiene, particularly around the time of birth, is vital to prevent this type of infection. Routine dressing of the navel with an antiseptic or antibiotic preparation is often carried out soon after the umbilical cord has broken.

'Sleepers' A variety of bacterial and viral infections may occur soon after birth; the bacteria or other agents enter the blood and cause generalised infection, as well as specific symptoms associated with particular infectious agents. For example, foals infected with Actinobacillus – affecting the kidneys – typically show sleepiness together with diarrhoea, kidney damage and convulsions. Other foals may develop signs of meningitis, which may also be due to other bacteria, and hepatitis – due to herpes (rhinopneumonitis) virus infection – may also cause similar signs. Poor intake of colostrum and bad hygiene are important contributory causes of this type of disease.

Umbilical hernia: Hernias at the navel may be present at birth; alternatively they may not be discovered until about six weeks of age. The hernia is simply a defect in the body wall where the two sides of the abdomen meet in the midline. Abdominal fat or occasionally part of the bowel (or, on very rare occasions, other organs) may protrude through the opening, under the skin. In most cases there will be a soft swelling which can be gently pushed back into the abdomen through the hernia ring. There is normally no pain or heat, unlike an abscess. The majority will require surgical correction in order to avoid the risk of the contents of the hernia becoming trapped or 'strangulated'.

ORPHAN FOALS

If a mare dies during or shortly after the birth of the foal, the orphan will either have to be fed artificially or be found a foster mother. It is essential that a source of colostrum is found in order to give the foal immunity from disease – colostrum should be given within twelve hours of birth, though if this is impossible, plasma can be given intravenously if available.

Finding a foster mother is the best solution for an orphan, a mare which has lost its own foal but still has milk – *very* occasionally, a mare will tolerate two foals. When first introducing the orphan to her, it is helpful to drape it in the skin of the dead foal or in the amnion and placenta. The mare and foal should not be left alone until it is certain that she has accepted it. Putting an ointment with a strong smell into the mare's nostrils or restraining her with a twitch may be useful, and tranquilisers or blindfolds are also used occasionally.

If no mare is available it may be possible to use a goat, or the foal may have to be bottle fed. Artificial mare's milk is available commercially, and for the first two or three weeks the foal should be fed every two to three hours with about 250–500ml. Training the foal to drink from a bucket as soon as possible is obviously a good idea, and it can then start on rolled oats and good quality hay. Even if being fed artificially, it is probably good to provide a companion (such as a good-tempered pony or a goat) and the foal should be put with a group as soon as possible so it can learn to socialise.

THE BEREAVED MARE

A mare which has lost her foal may show disturbed behaviour for some time. If the right combination of circumstances occurs, fostering an orphan on to her may be the best solution; but when no foal is available she will probably need attention, especially to prevent problems with the udder. Milk secretion will reduce quickly, but it is necessary to watch out for excess distension of the udder and if necessary draw off a little milk by wiping the udder with a damp cloth or squeezing the teat. The mare should also be watched carefully for signs of mastitis.

NUTRITION FOR MARES AND FOALS

In general mares should be kept reasonably fit and slightly on the lean side; they should certainly not be allowed to get fat. There is no need for any special diet prior to mating, and normal good quality feeding is all that is necessary; and in the first eight months of pregnancy normal rations are sufficient for most horses, other than those which are still being expected to work hard. Over the last three months, however, the foal is developing more rapidly and gets much larger, so an increase of about one third more than normal will be necessary, which should be phased in over a period of a few weeks. Shortly before delivery the requirements will have gone up to almost 50 per cent over normal.

In the last part of pregnancy the mare's protein requirements will be high, and levels should be 40–50 per cent above normal; good grass and other green foods, bran

A good mother encourages her foal to suckle and protects him from other inquisitive and maybe aggressive animals, including horses.

and oilseed cake for example, all provide high levels, but in many ways a commercial diet is often the most convenient way to provide for the mare's needs Sterilised bone meal for calcium should also be provided, approximately 90-180 grammes per day depending on size.

In the later stages of pregnancy she should be given more frequent, smaller feeds; she should have constant access to clean water throughout pregnancy. Deficiencies in the mare's feeding are, within reason, not passed on to the foal, because the foal takes whatever it needs from the mother and leaves her to make up the deficits.

In the first six months of pregnancy light work is perfectly alright, but after that it is probably best to allow the mare to do as she likes on good pasture rather than continuing with work.

Stallions do not appear to have any special nutritional needs, but it is important that food is of good quality and that vitamin and mineral levels are not subnormal.

Lactation
The demands of lactation are considerable, and begin on the day of birth – suddenly the mare has to provide for the foal's continued growth, plus enough energy for its exercise. Heavy breeds produce up to twenty litres of milk per day at the peak of lactation. The mare will need more water than usual during lactation, and may drink anything up to forty-five litres per day. Protein levels in the diet must be kept high (approximately 25 per cent of the total dry matter), and levels of fat-soluble vitamins (A, D, E) and calcium should be increased.

Foals reach twice their birth weight by about five weeks, and although lactation may continue for five months, solid food is usually introduced by about three weeks. The composition of the ration used is not very important, however, because the foal's main nutritional needs are still being met by the mare's milk. Commercial diets are probably the most useful way of feeding the young foal, but mare and foal should have access to grass and fresh water at all times; natural weaning continues up to the age of at least four months.

Orphan foals can be raised on recipes using modifications to cow's milk, or on proprietary milk substitutes, though some of these may need to be more diluted than is recommended in order to avoid constipation. At three days old, warmed water should be given at about 2½ litres daily, rising to 10 litres at five weeks – but it is important to take care that the foal does not take to drinking water in preference to the milk substitute. The fibre provided by good quality hay is essential for normal development of the bowel, so hay plus grain or cubes are introduced at two to three weeks. All changes should take place gradually, in the same way as natural weaning and feeding patterns, so that the bacteria in the bowel responsible for digestion have time to adapt to the new foodstuffs. Orphans which have diarrhoea should stop the milk substitute and be given glucose and water (50g per 500ml) for 24/48 hours; the milk substitute can be gradually reintroduced when the diarrhoea has settled down. The foal needs about 100ml per kg bodyweight per day, so a 100kg foal would need about 10 litres of milk substitute daily. Feeds should be small and frequent to avoid overgrowth of bacteria in the bowel and hence diarrhoea. Good hygiene in food preparation is very important, and adequate exercise is essential – it is important to see that an orphan foal does not become overweight.

At weaning, the foal's diet should contain around 16 per cent protein, with 0.8 per cent calcium and 0.55 per cent phosphorus; yearlings should have about 12 per cent protein with 0.5 per cent calcium and 0.35 per cent phosphorus. This can be achieved with many different combinations of foods, but in general a commercial diet is most practical.

WORMING AND VACCINATION

Vaccinations for tetanus and flu usually start when the foal is three months old, with a second injection four to six weeks later. A further booster for flu is usually given after another six months, and a booster for flu and tetanus is usually given one year after the initial course.

Earlier vaccination of foals should only be carried out if the foal has not received colostrum or if the mare was unvaccinated. Otherwise immunity inherited from the mother will make the vaccination ineffective. The mare should have a booster vaccination one month before foaling.

It is essential to instil obedience from early days. This youngster is leading in hand enthusiastically and co-operatively.

You should worm foals from 6 weeks old, unless diarrhoea due to Strongyloides makes it necessary to begin sooner. Dosing should then continue every four to six weeks.

The mare should have a routine worm dose every four to six weeks, and she should be dosed two to three days before turning her out to pasture with the foal. Preferably the mare and foal should be turned out onto clean pasture (clean pasture means not having had horses grazed on it for at least one year); if clean pasture is not available it is best not to turn the mare and foal out until June using other surfaced facilities instead. After turning out, the mare and foal should continue with dosing every four to six weeks.

Important points to remember in controlling infection with large or small strongyles are that:
i. The spring and early summer are the main periods when eggs are shed onto the pasture.
ii. The eggs develop into infective larvae over the next few months (May to October).
iii. Some infective larvae can survive on pasture right through the winter until May of the following year. This is the reason for trying to avoid turning youngstock out onto contaminated pasture before June.

7 RIDING AND SCHOOLING

by Sylvia Loch

All art, and every science, has its own principles and natural constraints. Nature itself follows certain strict laws, so it would be both arrogant and ignorant to assume that riding – a pursuit which arguably brings us closer to Nature than any other – isn't also governed by similar laws. Whether we consider riding as an art, a science or a sport doesn't really matter too much; what *is* important is to recognise the necessity of firm foundations in adherence to Nature's own laws; it is on these that we should build a correct method for all our equestrian activities.

The unique factor about horses and riding which has fascinated mankind since the dawning of civilisation and continues to do so today is the phenomenon of man and animal moulding themselves together into one being, one entity. The concept of the centaur, man and horse joined together, can only be achieved through communication, co-operation and a general sensitivity one towards the other. Few activities involve such conscious introspection and such a level of self-awareness. Thus for the serious rider, novice or experienced alike, there is an ever-constant quest for greater understanding. The most wonderful thing about riding is that you never stop learning. The very unpredictability of that beautiful, generous creature, the horse, makes the riding of him so challenging; yet this must not blind us to seeking a proper method. As long as we recognise this, our association with the horse will become more fulfilling, and more positively rewarding every day.

THE MENTAL MAKE-UP OF THE HORSE

Things are not always what they seem . . . Long before we put our foot in the stirrup iron, we should stand back and appreciate the horse for what he is. A charming domesticated animal quietly munching his hay in his box, the newcomer may be forgiven for assuming that he is there at our disposal, his only duty to behave as we wish. What is less forgivable is the assumption that when things go wrong, it is the horse's fault. 'Stupid!' 'Obstinate!' 'Unco-operative!' are some of the things we hear him called often from the most well-meaning riders. Yet few horses are naturally stupid, obstinate or unco-operative, and if these characteristics are displayed, it is generally through a lack of understanding on *our* part.

We must never forget that the horse is first and foremost a creature of flight. In his natural state he is a herd animal, a gentle social herbivore whose way of life is spent browsing lazily from pasture to pasture, but whose safety lies in swift flight. His excellent hearing and swift reactions therefore make him naturally alert to all possible dangers and potential predators. In the wild state, this beautiful animal only feels secure when he has familiar surroundings and sufficient space around him in which to feed (and to flee, if necessary); in addition, he thrives on having constant company, and the reassurance and stimulus – although not too close – of his own peers.

Today's riding horse is rarely able to enjoy such a lifestyle. It is therefore important that the caring owner provides another form of security to replace that which he has lost. A comfortable box or field shelter should be provided, and regular feeding, grooming and cleaning hours kept, so that a firm routine is established from which the horse will gain confidence and a sense of continuity. It is equally important to the behaviour pattern of the horse to provide as much company as possible. Human company can be as satisfying as equine to the stable-kept horse, provided there is plenty of it. A happy horse is one who is in constant touch with what is going on around him; in a large, busy yard with boxes facing each other, this is not difficult. For the horse kept singly, the ideal location is close to the approach of the house where the horse can see all the comings and goings. It is a grave mistake to shut a horse away in a box far removed from human habitation or active sights and sounds; even if he is being visited three times a day – this is not sufficient.

As for the horse's natural instincts to flee, these will gradually be diverted through proper progressive training, but it must never be forgotten that even the fully schooled horse with complete confidence in his rider can still be frightened by the unaccustomed or the unknown. To punish him when he shies, therefore, is a rash and foolish thing to do; better to soothe hm past the object of concern. A wise trainer will always seek to reassure, and this type of handling generally results in a horse which reacts less and less fearfully to strange sights and sounds. On the other hand, a horse which is frequently punished for shying or showing signs of fear is likely to grow more and more nervous, which may result in permanent nappiness. It takes a discerning rider to know when the horse is genuinely worried, or when he is just playing up.

Why do we school?

The proof of good schooling, and the level of achievement to which every rider should aspire, is when the horse can comply – physically and mentally – to all reasonable requests as and when they are demanded. For some the goals will be higher than others, but it is the duty of every rider both to himself and his horse to be in control. The successful trainer is not merely the one who produces a fit and supple horse, but one who has also learned how to rechannel all the instinctive resources of his horse – its strength, speed, sensitivity and impulsion – and can harness these to his own requirements. Real submission is when the horse offers this happily, rather than under compulsion; and real success is when it is obvious, from the look in his eye and his bearing, that he takes a pride in his own achievement.

The road to successful training, however, does not happen overnight. It is a long and painstaking business, and none of it will be lasting and beneficial if the rider expects too much too soon. Little and often is the maxim of the wise trainer; and from the outset it is important that he is able to make his wishes clear to his horse.

Schooling through reward

The language which all horses understand from the very onset of handling is that of reward and punishment. This concept is as important and relevant today as it was over two thousand years ago when Xenophon, a Greek cavalry leader and philosopher, wrote '. . . the art of equitation is based on rewards and punishments'. No truer words were ever written; but it is unfortunate that today too many riders, even amongst a nation of confessed horse-lovers, pay less attention to reward than they do to punishment.

In Xenophon's eyes, punishment was not so much the abuse of the horse, but rather the cessation or withdrawal of reward. For example, use of the positive seat and legs is often all that is needed to push a horse forward, and if a tap with the stick is necessary, then it should be applied immediately. Often just a sharp word to counteract deliberate naughtiness is wonderfully effective – the point being that if the horse is used to working in a state of calm reassurance with ready reward when he does well, he is more likely to respond quickly when his rider gives even the slightest indication that all is not well.

Too often, however, horses are schooled in a permanent state of discomfort with busy, punitive hands and over-use of the spur at every stride. Such horses will eventually lose that generosity of spirit which makes the teaching of them so much easier, and the trainer will make his task very much more difficult for himself.

Of course no horse is perfect, and there are all too many obstinate, recalcitrant horses which someone else has spoilt. These will need to learn all over again, but even with the most difficult pupil the application of pain rarely achieves very much, whereas psychology, quiet and consistent riding, encouragement and patience usually bring considerable progress. In other words, the rider says to his

RIDING AND SCHOOLING

Mental make-up of the horse
Why do we school?
Schooling through reward

MUSCULAR DEVELOPMENT

As trust grows, so should the horse's muscular development. Nothing must be forced, and exercise should be progressive. Moreover riders must realise that however pleasurable it is to gallop about, fast work on its own does not build up muscles – in fact it can do the reverse if it is not part of a logical programme. Conditioning and building up the muscles of the neck, back and rump will only be accomplished with slow systematic work over different kinds of terrain, combined with suppling exercises on the flat. Circle work on the lunge, on the longreins or under saddle will not only help the horse to turn and bend more easily, it will also play a part in combating the horse's own natural crookedness.

Once the muscles have developed along the horse's top line and become more pliant, it is easier for the horse to bend and flex his legs through supple joints. People talk glibly of 'engagement of the hocks', but do not always appreciate that the horse must step further underneath his body with his hindlegs, which in turn creates more impulsion and drive under saddle. Neither do they always appreciate that this ability is far from automatic, the easing and suppling process takes time, but the end result is a horse that can respond swiftly and easily to the rider's aids.

Too many riders fail to understand that their horse *cannot* always do what they ask of him, because he is physically incapable. Certainly some horses may be wilfully disobedient and require firm handling, yet it is far more often the case that when a horse refuses a jump or is unable to execute a turn in the correct manner, he is physically uncomfortable and simply not up to what he is being asked to do with the rider's weight on top.

Balance is the cornerstone of good riding, and it takes time to mould, and educate and help the horse to rebalance himself in all movements and gaits when he is carrying a rider.

RIDING AND SCHOOLING

Defining reward
The importance of correct
    and efficient riding

EQUINE PRIDE

How many people are aware that the horse enjoys a strong sense of pride? Horses love to be clean and smart; when they sparkle outside they soon begin to sparkle within. Every healthy happy horse will show off when he has been well turned out and plaited up for a show or a day's hunting. Like human beings, the horse has a tremendous sense of occasion. You need only look at the faces of the Queen's Birthday Parade horses as they jog in procession past the eager crowds on that special June morning in London, and sense the pride they take in their task; and this is particularly so if in the company of others. Admiration and attention are features of equine life which are as important as good husbandry. Yet many people fail to realise that if they encourage this natural pride, even the most common task will be improved, for the majority of horses really do seek to please.

horse 'If you do this for me, I will reward you; if you don't, you won't get any reward; and if you do *that* (something unacceptable), I shall make life very uncomfortable for you as I am prepared to perservere all day if I have to!'

An eighteenth century English Master, Philip Astley, who trained horses for the High School, the cavalry and the circus, wrote in a tribute to the horse at the beginning of his book *Equestrian Education* '. . . Such is the greatness of his [the horse's] obedience, that he appears to consult nothing but how he shall best please, and if possible anticipate what his master wishes and requires . . .'. If more riders quoted this at the start of every schooling session, they might be pleasantly surprised at the results achieved from such a positive attitude.

Defining reward

There is every difference between sentimental sloppiness, and disciplined training through reward and punishment. The first rule is always to reward or punish *instantly* so that the horse knows why he is being thanked or rebuked. There are many ways of rewarding; the obvious ones are a word of commendation and praise, a stroke with the hand, or an occasional timely sugar lump. Another reward, much overlooked nowadays, is the yielding of the hand when the horse is nicely on the bit and going forward as we wish him to. This means a very slight easing of the contact of the reins for a few strides, to help the horse relax through the jaw whilst still maintaining the rhythm of the gait. A more obvious way is to let him stretch his neck completely and go forward on a long rein, although this should be done at the end of an exercise, not during.

It is not of great relevance which reward is used, or in which sequence; what is important is that the horse truly *does* recognise that he has done well. He has an excellent memory, so this is by far the most effective way to train any horse, and in particular the 'unmade' youngster. It should also be carried through to the highest levels of training, for even Grand Prix dressage horses still need praise and reward if they are to give of their best.

THE IMPORTANCE OF CORRECT AND EFFICIENT RIDING

Today, insurance companies and the medical profession consider riding to be a high risk sport. In terms of actual numbers of serious and fatal accidents, it takes pole position, above even motor racing, sailing, skiing and flying. Since our changing environment provides fewer and fewer quiet rural areas in which to ride, riders today have a duty to themselves, their horses and the general public to ride more proficiently and with greater control than ever before. Unfortunately the general standard of riders on our country lanes and highways shows that this is not always the case.

Generally, the education or training of horse and rider may be divided into roughly three categories:

1 basic or preliminary
2 secondary
3 advanced or superior, leading to specialisation.

Too many riders, however, seem content to confine themselves to a stage which does not approach even the first category. Having failed to recognise that to be safe on horseback, rather more is required of them than just to 'stick on' at walk, trot, canter, gallop and even over jumps, they themselves become a liability to others. To be no more than a passenger does not constitute real riding; probably the best guide for a rider who wishes to prove himself in the basic stage of equitation can be found on the back of any BHS novice dressage test sheet. The word 'dressage' could well be replaced with the word 'horsemanship', for whatever the discipline, this clearly defines an acceptable standard for all horses and riders in their basic education. Moreover, any rider who takes his horse and

An all too common sight. The generous spirit of the horse can soon be abused by riding like this. So-called schooling on a heavy rein contact with the horse's head visibly pulled down by low-set, clenched hands can produce incorrect flexion in the neck and a painful back. Submission may be achieved but at the expense of the horse's pride and gaiety.

himself out alone on the road without having reached this standard, is putting himself and others at risk. Unfortunately this is too often the case, which only helps to swell the statistics of serious and fatal riding accidents.

Real control

Whilst all this may sound very academic, none of these requirements should be beyond the capability of any rider who is considering buying his or her own horse. Later on, some understanding of collection (secondary stage), and an appreciation of its benefits in keeping a horse under more control in difficult conditions, will also prove invaluable. Riders should be prepared to invest the necessary money in lessons given by qualified people who can help them in this direction. The investment will pay dividends, because increased safety and control – which, after all, may ultimately save life and limb – is beyond price.

We must remember that not so very long ago, no gentleman or lady would consider riding out on a hack or hunter which was not schooled to basic collection; it was simply 'not done'. However, in those days there were plenty of knowledgeable grooms and professional horse people around who undertook the schooling and ensured a steady supply of educated riding horses. Unfortunately this is not the case today. Ironically, when roads are at their most hazardous, and despite the popularity of dressage, many young horses are broken in badly and spoiled by people of little experience. In addition, there is still a hard core of people who insist that schooling is a waste of time, and that such niceties have no place in the framework of general purpose riding. How wrong they are!

Soundness

There is another argument in favour of further schooling, and one which shows how important it is to teach the horse and rider a degree of self-balance: it concerns the matter of soundness. All the great cavalry instructors of ancient and modern times have recognised that horses ridden permanently on the forehand were more likely to break down. Cavalry schools were therefore set up to teach young officers and recruits the basic rules of balanced riding. Today such truths are often forgotten in the soft sand of the indoor riding school, and it is only when riders take their horses out into the real world of tarmac roads, rocky paths, plough and marsh, that they discover the error of their ways. Furthermore sustained trotting on hard surfaces, allowing most of the horse's weight to be

FEI REQUIREMENTS FOR HORSE AND RIDER AT NOVICE LEVEL

The Rider:
Adopts a correct position in the saddle with a seat *independent of the reins*. Is free from stiffness in any part of the body or limbs;
Maintains a constant, *light, acceptable* contact with the horse's mouth;
Applies simple aids *clearly and effectively, avoiding all unnecessary movement*.

The Horse:
Has true, regular and unhurried natural paces;
Is calm, relaxed and obedient to the aids of the rider;
Maintains an improved natural outline, balance and rhythm;
Moves freely forward, without collection but with active hindquarters;
Accepts the bit willingly and without tension or resistance;
Remains straight when moving on straight lines and bends accordingly when moving on curved lines;
Executes transitions smoothly and remains still when halted.

[The italics are mine, included here for emphasis]

RIDING AND SCHOOLING

Preparation of the rider
A safe secure seat
FEI requirements

carried on the forehand, will have serious results for the majority of horses. Constant jarring of the bones and joints will take its toll with the toughest of mounts – yet many cases of chronic lameness could be avoided if owners understood better the importance of correct schooling.

Let us now look, therefore, to preparing the rider for the levels discussed above, before allowing him loose with a horse. It is particularly important for a rider to have achieved this standard of competence if he wishes to 'bring on' a young, inexperienced horse since it is generally agreed that there is no greater mistake than allowing the blind to lead the blind. Of course there will always be the success stories where a completely 'green' horse has been nicely brought on by a 'green' rider, the one apparently teaching the other through trial and error along the way. There are also the horror stories. Sometimes the horse is ruined through general ignorance and lack of understanding; on other occasions it is the rider who finds himself increasingly overfaced by the problems encountered with an intractable pupil, and confidence is irrevocably lost. At worst, accidents ensue, for there is nothing so bad for the young horse as a trainer who is either afraid or nervous of him; lack of confidence in one tends to spark off a chain reaction in the other. All this is best avoided, by only ever putting the rider in a situation where not only does he understand exactly what he is doing, but is physically capable of achieving good results.

PREPARATION OF THE RIDER

A safe secure seat

The foundation of all good riding, irrespective of the discipline in which the rider may later specialise, is the acquisition of a solid base of support in the saddle: in short, the rider's seat. Only when he can achieve a deep, still and supple position which remains in contact with the saddle at all gaits, will he be able to change the emphasis of his seat – safely – for different purposes. At one extreme this might be the extreme forward position of the flat race jockey in the final sprint past the post; at the other, the accentuated depth of the High School rider of Vienna or Saumur,

A proud horse usually has a proud rider. This rider is at one with his horse who, through correct schooling, is able to carry himself and his rider in comfort from a supple, rounded back and a relaxed jaw.

All may go well until the horse makes a sudden stop, when the lack of balance through the torso and muscular tension in the buttocks, thighs and calves will cause this to happen.

as he collects his horse into a stately passage round the perimeter of an ornate indoor riding hall.

The concept of the basic or classical seat is surprisingly simple yet for years, even centuries, it has been the subject of hot debate, and whole sections of books (often differing vastly) have been devoted to the subject. It may help us to define our priorities however, if we look at it from the horse's point of view.

Most people recognise that the horse's back is at its weakest just behind the saddle, *ie* at the loins. This part covers the horse's kidneys and is not designed to carry weight, so it is a thoughtless rider who slaps his saddle too far back on the horse's spine at this point. The strongest part of the horse's back is just behind the withers, and his centre of activity is roughly under the centre of the saddle. In order for the rider to sit in harmony with his horse, therefore, it makes sense not to slide to the rear of the saddle, but to keep the weight central in the saddle, with the rider's own centre or solar plexus as close as possible to that of the horse.

This is called *riding in balance,* and provided the rider maintains this hold or centre, he will always be at one with his horse and remain over the strongest part of the horse's back.

The main threat to a good position in the saddle is rider tension, and in the average riding lesson, not enough emphasis is placed on the importance of a relaxed but supported stomach. Since the solar plexus is situated in the pit of the stomach, tension here will spread throughout the nervous system to the rider's chest, back, neck, shoulders, arms and legs. This explains why so many beginners and nervous riders take up a cramped position in the saddle, immediately inclining them away from the vertical. Moreover tension has a spiralling effect, and once the horse is in motion the thighs and buttocks will also tense, thus pushing the seat up and away from the saddle. The result of all this is inevitable: a sudden stop or jerk from the horse will immediately cause the rider to fly forward and/or fall.

It is of little use telling beginners to sit correctly. What must be explained is the importance of relaxing the stomach, so that the seat can be opened and widened to form as broad a base as possible for the upper body. Once this is achieved, the rider must then stay erect in his upper body by growing tall and proud through the torso. The idea is to be as upright and natural in the upper body as if you were standing erect on the ground, and as relaxed and unconstricted from below the waist as possible, thus allowing the weight to flow downwards and unite the rider to his horse.

Correct saddling. Unfortunately, many saddles used for novice riders are not suitable for acquiring a good seat. A rider cannot be taught to sit centrally if the lowest part of the saddle is too far back, for gravity will inevitably slide him towards the horse's loins. To avoid this, some schools which cannot afford suitable saddles will place a thick piece of foam rubber under the cantle in an attempt to rebalance it; but there is no substitute for a correctly cut saddle where the lowest point must be *central* as seen above.

The independent seat

After approximately ten voltige sessions on a blanket (see p.164) a rider's seat is normally improved out of all recognition. Whatever other faults he/she may have, these will be much easier to deal with once a truly independent seat has been attained, at all gaits and over small jumps. Once the seat is deep, it gives the rider the appearance of being 'glued to the saddle'. Taken point by point, the following characteristics should be unobtrusively obvious:

- a snug, all-round contact of the total seat, with thighs, fork, seatbones and buttocks adhesive from front to back;
- a proud, erect and open position of the torso with shoulders held comfortably, but not exaggeratedly, back and down;
- a supple, flexible lower back or loin which is capable of supporting the torso and flexing forward to absorb the horse's movement;
- bent elbows and relaxed forearms with the hands held comfortably and lightly just in front of the waist;
- the rider's stomach incurved a little forward and down; the rider makes an imaginary attempt to allow the waist to meet the hands;
- the thighs lie flat against the horse's sides, unconstricted by the round thigh muscles which are pushed behind the thigh bone;
- the knees remain as low as the stirrup leather allows against the saddle flap, neither turned away nor pressing tightly inwards;
- the lower leg lies relaxed and light against the horse's side, close enough to apply the aids, but never actively gripping;
- the foot faces in the same direction as the horse, the ball of the foot resting on the stirrup and the heel dropping a little lower from a relaxed ankle joint.

If a line were taken from the top of the rider's head, through his ear, through the point of shoulder and hip to the heel, it would be roughly perpendicular.

Once an independent seat is achieved both on and off the lunge, the handling of the reins and the application of the aids become very much easier. It is also then a simple matter to adapt the seat for different purposes, as we shall discuss later.

Mounting

Most people who have achieved a good seat will be good at mounting, since an important part of the broad-based open seat is suppleness in the hip joints, and this of course helps the rider when he/she has to swing the leg over a very tall horse.

Nevertheless, it is much better both for the horse (and for the saddle) to mount from a mounting block if any trouble is experienced; mounting is never a very pleasurable experience for the horse, and whilst some riders can jump up easily and lightly causing minimum discomfort, others cannot. Inability to mount correctly can cause one-sidedness in the horse and even provoke back problems; it can also twist the tree of a valuable saddle.

Let us therefore just recap on the absolute essentials for the procedure of mounting; but before we go through the mechanics, phase by phase, let me remind you first of all to pause a moment to greet, stroke and reassure the horse. An all too common and insensitive sight is the rider who has his horse brought out by a groom and who clambers on without the horse even knowing who it is he is about to carry. A word of greeting first is most important. Then the rider should:

1. Take the reins lightly in the left hand which he then places just in front of the pommel (if a double bridle is used, the curb rein should be held even more lightly than the snaffle rein);

2. Face the tail of the horse and, turning the iron towards you in a clockwise direction, place the left foot well into the stirrup iron so that he feels a firm hold under his foot;

3. Place the right hand lightly on the seat of the saddle, *not* to pull himself up, but merely as a means of support;

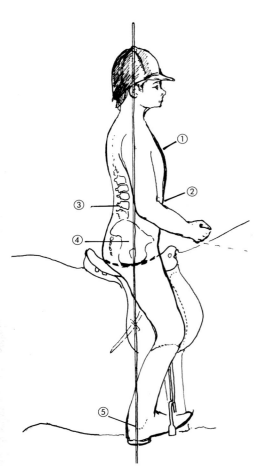

From a central, balanced or classical position which shows a perpendicular gravity line through the head, shoulder, hip and ankle, the rider can learn to sit safely and at one with his horse. From this upright posture, other positions such as the jockey crouch or jumping seat, which require a greater degree of compression from shortened stirrups, will be much easier to achieve.

1 expanded, open chest
2 waist not pulled in
3 lower spine with natural incurve at loin
4 pelvis roughly upright with *all round* contact in lowest part of saddle
5 supple ankle joint although here the heel is a little too deep

Mounting
1 Prepare!
2 Push off!
3 Swing into position!

Dismounting
4 Both feet out of the stirrups, and swing position!
5 Lean into the horse and drop!
(For full explanation, see text)

1

2

4

5

4. Spring off the ball of the right foot so that the right leg can be swept clear over the horse's back. This is facilitated by leaning into the horse during the manoeuvre;

5. Before lowering himself into the saddle, the rider should place his right foot in the right stirrup and deliberately open his thighs before dropping his weight down quietly into the centre of the saddle.

This last stage is very important, since it opens the seat and gives a broad base of support in the saddle. Too many people sit down firmly on their horse with closed or 'clothes-peg' thighs, which never allows maximum adhesion of the seat. Having taken up the reins at the required length make sure that you are comfortable, deep and wide in the saddle before asking the horse to walk on. To maintain a good seat, this final settling must never be taken for granted and it is worth pausing a moment to make any final adjustment, such as pulling away the flesh from under the thighs to allow the thighbone to lie flatly, as already discussed.

Dismounting
Dismounting is altogether simpler. The rider should:
1. Halt his horse quietly, and relax the contact;
2. Take both feet out of the stirrups, and take the reins in the left hand, which rests just in front of the pommel;
3. Support the right hand on the withers or take a handful of mane and, leaning forward slightly, swing the right leg backwards and clear of the horse's back and the saddle;
4. Alight on the ground, with the back towards the horse's head as in mounting;
5. Slacken off the girth, slide the stirrups up to the top of their leathers and tuck in the slack, before taking the reins over the horse's head, to lead the horse in the normal way.

THE AIDS

The true meaning of the word 'to aid' is sometimes forgotten in the average riding lesson. To 'aid' is to *help* and in many ways it is regrettable that the old word *helpes* which we find in both 16th- and 17th-century equitation books is no longer in usage. Too many people translate the aids as commands, and forget that these are supposed to help the horse not only to understand what we want from him, but also to carry it out – with as little interference as possible.

The aids are categorised into two sections: natural and artificial; and only when the rider has a good seat will he be in a position to use these correctly. The natural aids are: the legs; the body; the hands; the voice.

The artificial aids consist of stick and spur, and include the three different types of martingale (running, standing and Irish), and the various nosebands such as the drop, flash and grakle.

None of the aids will be truly effective if the rider has not achieved a firm deep seat in the saddle. The proof of good riding is to see nothing, or at least very little, happening between rider and horse; ideally they should move effortlessly and as one body from movement to movement, and this can only be achieved when the natural aids are quietly and discreetly given, and without the use of artificial aids.

Let us look briefly at the application of the different aids.

Legs
The legs have four purposes when we ride:
1. to aid security and balance (mainly the thigh and knee);
2. to create impulsion and energy in the horse (mainly the lower leg);
3. to support and direct the horse in turns (mainly the lower leg);
4. to control the hindquarters (mainly the lower leg).

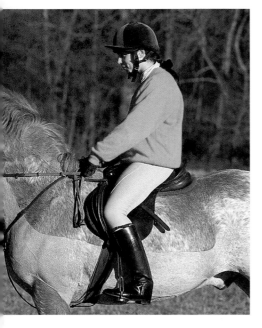

The instinctive position of most beginners is to collapse at the waist, drop the chest, round the shoulders and lose the support of the back.

Nowadays we hear a great deal about driving legs, but these can do more harm than good. In the classical school – which all riders would do well to emulate – the leg is required to be effective through its lightness: a light, asking leg should quicken the horse's responses and create impulsion, and the most effective spot for its use is just behind the girth. Short, quick squeezes given in rapid succession with the lower leg (taking care not to alter the position of the seat and thigh) will achieve very much more than repeated booting in the ribs, which annoys the horse and eventually deadens his sides.

As well as creating the energy to make the horse go forward, the legs also direct and support the horse in turns, and guide and control the hindquarters. Thus in a circle to the right, for example, the inside or right leg will support the horse on the girth – giving the forehand something to bend around – whilst the outside or left leg will move quietly behind the girth, easing the horse into the turn and preventing the hindquarters from straying outwards.

The role of the supporting leg on the turn must never be underrated; it is wonderfully effective in teaching the circle to the young horse as it helps to supple and bend him. In turns and circles, there should be a little more weight into the inside leg, ie in the direction in which the horse is travelling – although this useful aid will become totally ineffectual if the rider does not maintain his balanced seat; and he must guard against his seat slipping to the outside when he uses the inside leg – a common fault.

Body

This includes the entire torso and the seat, which we have already discussed in some detail. Too many instructors tell their riders to relax their upper bodies, and this can have serious results: it is better to ask them to sit tall and proud, stretching upwards through the abdomen and chest (rather than through the back, which tends to raise the shoulders). The importance of a supple back must also be emphasised and this can be done on the ground, explaining to the rider how the small of the back or loin (which naturally curves inward in the standing figure) acts as a spring, absorbing the movement under saddle. It is this part of the body, together with the joints, which unites the rider with the movement of the horse thus preventing bumping and jarring of the pelvis and spine.

At all times the rider's head and body should face the direction of the movement, thus his gaze should follow through between and beyond the horse's ears. A rider must learn to turn his upper body from the waist; this is an important aid in turns and lateral movements. In a simple circle to the right, for example, the rider's shoulders should turn slightly right as the horse turns his shoulders to the right. The same principle is applied in the shoulder-in: many riders have difficulty in teaching their horses the shoulder-in, yet more often than not the problem lies in the fact that the rider has forgotten to turn his own shoulders in and look in the direction of the movement.

Because the spine controls the pelvic girdle and ultimately the legs, the back plays a greater part on the position of the rider's seat in the saddle than most people realise. Different effects are therefore made on the horse's back, depending upon the way in which the rider uses his spine. These can be detrimental or helpful to the horse. The effect of the back on the rider's seat is described in the inset box. These are sometimes referred to as weight aids, but they should not be knowingly practised until the rider is very well established in his basic seat and controlled posture as described on p.166.

The hands

Through the reins and the bit, the hands are the means by which we communicate our intentions to the horse, ie explaining where we wish to go and in what manner. Hands should never become dead weights on the horse's mouth, and to keep the wrists supple it is helpful to imagine a polite conversation existing between the

THE EFFECT OF THE BACK ON RIDER POSITION

● A rounded back or spine has the effect of putting the rider on his tail bone instead of allowing his weight to settle down onto the triangular shaped ridges of the pelvic floor. Thus although this seat may appear light, it causes instability, and cannot be recommended.

● A spine which is flattened or braced out of its natural S-shape puts more pressure on the back of the seat bones, and this is very effective as a driving aid. Many riders develop this seat for all work, but if maintained, it deprives the body of spring in the lower back or loin, and can cause stiffness not only in the rider, but ultimately in the horse's back as well.

● If the S-shape of the spine is accentuated by hollowing the loin slightly forward, (waist towards the hands) it has the effect of increasing the downward pressure on the frontal ridges of the seat, and this is a useful aid for checking the horse, or achieving rounding and collection by alleviating pressure further back.

A LIGHT HAND

Not enough emphasis is put nowadays on the value of light, quiet hands. Light hands mean light horses, and the rider who has schooled his horse to respond to the gentlest of contacts and quiet leg and seat aids will not only look more impressive, but will enjoy all his riding so much better. Who wants to end their ride exhausted, having had a permanent battle with the horse? Yet some riders do this every day!

Normal snaffle bridle with cavesson noseband.

Normal snaffle bridle with drop noseband fitted correctly so as not to restrict horse's breathing.

rider's hands and the horse's mouth, the degree of contact depending on the horse's stage of training.

In the young horse, the hands should be held low, just in front of the wither, allowing the horse free forward movement from a gentle contact which follows the oscillations of the horse's head and neck without disturbing his balance or attempting to fix his head position. As the horse becomes more accustomed to this contact he will naturally become steadier in the head and neck, until eventually he should accept the bit with a pleasant moist mouth, neither leaning on the rider's hand nor losing the contact with it. In these early stages, it is for the rider to judge whether or not to ease the elbow slightly forward when the horse asks for more freedom, thus avoiding continual changes of rein length. In the mature horse which is ready to discover, or has already progressed to self-carriage and basic collection, the hand may be held a little higher (just above the wither) so that the rider's elbow remains against his side; any yielding can then be achieved through the fingers which establish a gentle music ('light sponging') between horse and rider. Tightly clenched hands are to be avoided as this creates tension in the arm, shoulders and upper body.

It is axiomatic that a rider with an insecure seat and hunched shoulders, for example, will have bad hands. The whole secret of good riding is to be in balance with the horse, but the unbalanced rider has no alternative but to use the reins from time to time as a means of support. Nine times out of ten, bad hands are caused by a faulty seat; however, there *are* other problems which may be the cause, and which may be easier to put right, such as tension in the neck and shoulder, or a stiff wrist which must learn to rotate and yield as well as to support the hand. Much can be achieved through exercises on these specific areas both on and off the horse.

To sum up, the aids of the hands are twofold: through the reins and the bit they:
- guide the forehand in the required direction; and
- regulate the impulsion and flow of energy created by the legs and the body, checking or allowing as required, but always in conjunction with the aids of the legs and body.

The aids of the hands should always be fractionally later than the body and leg aids. For example, many people stop their horses simply by pulling back on the reins; only when this fails to take effect, will they remember to close their legs and check with the seat. The correct way to halt is of course in reverse: close the legs, check with the seat by pushing down and forward whilst keeping the shoulders back, and close the hand. Never, ever pull back. This has the desired effect of pushing the horse forward into the halt, which keeps him together and balanced — rather than the undesirable effects of pulling him back into the halt which is unbalancing and often painful, especially for a horse with a weak back.

Rein control
Once the horse is going nicely forward, the rider's shoulders, arms elbows and hands should, to all outward appearances, remain as still as possible. Nevertheless the music of communication is there, a secret shared between horse and rider which may involve the sponging of the rider's fingers when he wishes to ask or to give, a slight easing perhaps of the elbow, to raise or lower the arm, a rotation or turning of the forearm from the elbow which will open the rein and indicate to the horse that he is to turn, and other minute requests. The happy horse plays or chews a little on his bit, and relaxes his jaw in answer to the asking or sponging of the fingers and the forward aids of seat and leg. When the rider feels his horse 'give', he should reward the horse by yielding the contact very slightly before asking or 'sponging' again, with the rider's hand apparently still to the onlooker.

Holding the reins
For normal riding, the reins should be held in both hands.

The Snaffle Bridle:

The rider should pick up the reins as though picking up a pencil from a table, then turn the hand so that the thumbs are uppermost and the knuckles face towards the horse's head. Whilst some riders prefer the rein to enter the hand under the little finger, the most popular and comfortable method is to hold the rein between the third and the little finger. The rein then passes over the inside of the fingers which curl themselves round the rein, and the slack comes out between the first finger and the thumb. Some riders like to secure the slack with a bent thumb, as they can exert more pressure on it; others (particularly advanced dressage riders) prefer to hold the thumb flat as this gives a lighter feeling in the hand.

The snaffle bridle was originally designed for the schooling of young, unmade horses, and as a racing bit. It was also the bit for stable boys and errand boys, the idea being that they would do less harm to the master's horses with a snaffle than with a curb. Nevertheless, whatever we put in our horses' mouths should be treated with respect. Even the snaffle can become an instrument of torture in rough hands, for it acts on the bars of the mouth which harbour a bunch of sensitive nerves. Provided, however, that the rider rides his horse forward into an allowing contact from a good basic seat, the snaffle will yield good results. Horses soon learn to accept this bit at all gaits and for all purposes *ie* work on the flat and over jumps. The action of the bit in conjunction with the aids of the legs and seat (as already discussed) is perfectly adequate for straightforward riding – provided, of course, that the horse's mouth has not been previously spoilt.

There are many different types of snaffle, and the thicker varieties are definitely preferable to the thin jointed snaffle which, in rough hands, can cause pain and pinching. Twisted and gag snaffles are positively punitive and are not to be recommended. The straight-barred rubber snaffle is the mildest, then the vulcanite (straight-bar); and often a horse which is giving difficulty in a jointed snaffle will become sweeter in the mouth if placed in one of these milder bits. It does not always follow that a severer bit will improve a horse which pulls; very often the opposite is the case.

It takes a very tactful rider to achieve a perfect headcarriage in the snaffle bridle. People should appreciate that the idea behind the snaffle is to encourage the horse to stretch his head and neck forward and take a reasonable contact for free, forward movement. It is, therefore, an excellent bit for bold cross-country riding. However, whilst the horse's head may be raised by the action of the snaffle, it was never designed to drop the horse's nose, encourage flexions or to place the head in a position approaching the vertical. If more people were made aware of this fact, there might be fewer struggles on horseback, and fewer horses ridden with a strong fixed contact – a very punishing way of riding. Only through correct work *behind* the saddle, *ie* engaging the hindquarters through suppling exercises, will a rider succeed in obtaining a good rounded outline in the snaffle – and this will take time and patience. It is a complete mistake ever to try to fix a horse's outline through the hands.

The Double Bridle:

At the beginning of the century, when riders prided themselves on light hands and the ability to handle their reins correctly, the double bridle was in common usage for every educated rider who rode a mature, educated horse. It was considered not only the safest but also the most suitable general purpose bridle for hunting, showing and hacking. It was also *de rigeur* for use in the cavalry and the mounted police force.

Rein control formed an important part of the average riding lesson, and riders learned to respect the double set of reins and understand the dual purpose of the bridle. They learned to 'give' or allow with the curb rein for general forward work, and to 'ask' with it in order to check, drop the horse's nose and allow him to flex from the poll for the more collected work. The action of the double reins was

Double bridle with reins held in traditional manner, encouraging the horse to flex at the poll and relax the jaw.

RIDING AND SCHOOLING

Artificial aids
Different disciplines and
styles of riding

invaluable for 'reining the horse in' without losing balance after a sustained gallop before a jump or a covert on the hunting field. It afforded absolute control without resorting to strong hand action, and all was achieved with minimum fuss.

Today it is the snaffle bridle which is popular, but all too often it is accompanied by every kind of artificial gadget to compensate for its limitations; many people – and not least, probably, a number of horses – rue the passing of the double bridle from so many equestrian circles.

With the double bridle, the reins are picked up in the same way as for the snaffle, with both reins passing across the inside of the fingers. When the hand is rotated into the normal position (as described before) the two reins are then separated either by the little finger or the fourth finger of each hand, with the snaffle rein crossing over the curb rein to the outside of the rider's hand to be manipulated by the little or fourth finger; the curb rein remains on the inside to be manipulated by the second or third.

It is vitally important that the rider learns to distinguish between the two reins, which should be held quietly and lightly.

ARTIFICIAL AIDS

The stick will vary in shape and size according to the task required of it. Generally a schooling whip will be considerably longer than a showing stick – there is no stipulation in the dressage rules as to the length, but between 20 and 36in is common. A hunting whip carries a thong and a lash. Whatever stick is used, it should be carried in the palm of the hand with the head protruding a few inches in front of the fingers and pointing towards the horse's opposite ear. The stick should never be used in front of the saddle; as a punishment its application should be seldom. There are many successful horses which have never ever felt the real sting of the stick on their flanks, only the occasional tap.

The stick must never be applied whilst holding the reins in the same hand. Instead, all the reins should be placed into one hand while the stick is applied, otherwise it will upset the horse's balance and probably jog his mouth. A well trained horse will not be afraid of the stick since generally it is only used as an extra aid to push him on; for him, it occurs as a sharp or gentle tap – as the case requires – applied behind the rider's boot. If a horse has shown real disobedience – as opposed to just high spirits – and a timely reminder is in order, then this must be done immediately, and once only should be all that is necessary.

The spur is another aid to encourage lightness, impulsion and precision. Few riders are expert enough to ride with rowels on their spurs, and unlike the stick, the spur should never, ever, be used as a punishment. Spurs should only be used by riders with a totally independent seat, who have complete control of their legs and will not inadvertently turn and grip with the lower leg thus digging the spur into their horse's side.

The spur should be used sparingly, the inside merely brushing the horse when required. The toe must never be turned out so that the back of the spur deliberately prods the horse (as is sometimes seen); the classical masters speak of the 'stroke of the spur' and the 'breath of the boot'. This indicates the degree of lightness with which the spur should be applied; it is there only as a precision aid – a finishing touch.

Some sensitive horses (especially those who may have been misused) positively shrink from the touch of the steel, so it should never be automatically assumed that a horse is to be ridden with spurs. It is preferable to school most horses *without* spurs for most of the time, and only to add them to emphasise some point, particularly if there is some teaching difficulty with a new movement. In the case of the more advanced dressage horse, spurs are a compulsory part of correct competition dress (after the elementary level), and should be worn from time to time to accustom the horse to their feel.

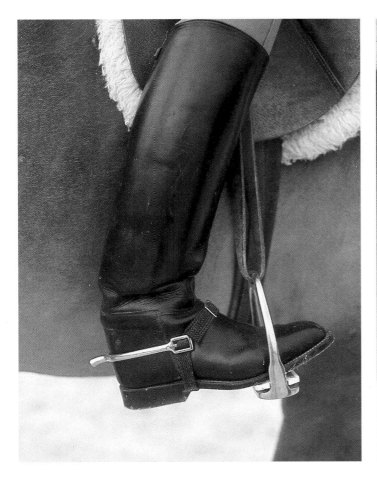

DIFFERENT DISCIPLINES AND STYLES OF RIDING

Once the rider has established a good independent position in the saddle at all gaits, and when he fully understands the use of the aids, as well as correct rein control, he should attempt to put as much mileage on his clock as possible before thinking too much about specialisation. There is nothing better than to grasp an all-round education in horsemastership, *before* settling for one particular discipline, and it should be remembered that a good foundation in dressage or basic schooling will inevitably improve the prowess of both horse and rider in every other discipline from showjumping to steeplechasing.

In recent years, dressage has fortunately begun to shake off the cloak of mystique which once surrounded it. It is the fastest growing riding discipline in the United Kingdom, and as cross-country, showjumping and event courses become more and more testing, so people are realising that as far as the most successful combinations are concerned dressage is already playing an established part.

Even in the hunting or polo fields – which have always prided themselves on a certain slap-dash approach – the benefits of a properly schooled horse are enormous. There is nothing more thrilling than to ride a fast, bold, exciting horse: but how very much more pleasurable the experience becomes when that fast, bold, exciting horse is also easily manoeuvrable, light to turn and absolutely in tune with his rider's wishes.

The Classical Dressage Seat

A good balanced riding seat (as discussed on p.166) is sometimes known as the classical seat, because it was first developed by the Greeks at a time when riding was beginning to be recognised as an art as well as a sport. In fact the word classical is not as daunting as it may sound. Generally, it denotes something which

A good jumping position. Note folded but supported back and torso, heels down, balanced angles of arm, trunk and legs. The hands are able to 'follow through' with a steady rein contact.

A bad jumping position. Note the rider is not in balance with the horse. The shoulders are stiff; the body has not folded; the legs are gripping so that the toes stick out; and the rider supports herself with a heavy rein contact instead of allowing her hands to follow through.

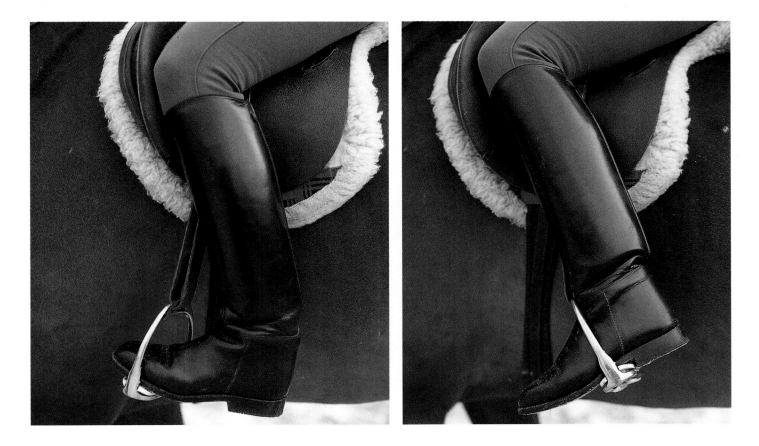

does not defy Nature's laws, which is practical yet aesthetically pleasing, and which is straight and true and never goes out of fashion. Whilst most people recognise that the classical seat is the perfect seat for dressage, it is not always appreciated that it is a good basis for every other seat.

If this sounds odd, let us now look at the forward or jumping seat.

The Jumping Seat

The forward seat as adopted for fast cross-country work at the canter or gallop and especially over jumps – is generally acclaimed to be the discovery of an Italian army officer Federico Caprilli, whose methods came into fashion at the beginning of the twentieth century. Before that, most of those people who raced or hunted leaned backwards over jumps, and it was only in 1897 that an American, Tod Sloane, brought the forward crouch of the modern jockey or point-to-point rider to Europe.

To start with, many people rejected the Forward Seat, glibly dubbing it 'the monkey seat', but fortunately the cavalry schools of Europe recognised it as a saving grace for the backs of thousands of long-suffering troop horses, who for too long had struggled with sloppy riders bumping on the back of their cantles in a seat which was neither classical nor forward.

Today we all take the forward seat over jumps for granted, but let us now discover how it differs and what it has in common with the classical seat as described on p.166.

The secret of a successful forward or jumping seat is the ability to remain in balance with the horse in motion, using closed angles of the body and limbs, the weight of the rider directed over and onto the stirrups, and the back of the seat rising clear of the saddle from a firm knee position. In order to achieve this the stirrup is considerable shortened, so the whole emphasis of the rider's body is shifted upward, inclined and forward, instead of downward, perpendicular and forward as for the classical seat.

For general purpose riding, this leg position is adequate, although ideally the stirrup leather should hang straight down with the lower leg further back. However, the ankle is supple, allowing the weight of the rider to flow downwards.

This leg position, if maintained, creates instability in the rider's feet. With the lower leg pulled up from a tense thigh, the downward flow of gravity is obviously restricted.

RIDING AND SCHOOLING

The general purpose seat
How to achieve the forward seat
Training the young horse

HOW TO ACHIEVE THE FORWARD SEAT

The following instructions for the correct jumping seat are taken direct from the works of Littauer, a prolific writer and teacher, and a former pupil of Caprilli; he was responsible for taking the forward seat to America and promoting its use all over the world. Apart from the new angles and 'tilt' forward in balance, many of the basic premises of the classical seat remain the same.

1. Place yourself in the saddle so that the crotch of the breeches touches the pommel;
2. Sit with pelvis tipped forward, not on the buttocks; push knees forward and down;
3. Draw the lower leg back so that the stirrup leathers are vertical or slightly behind the vertical;
4. Then the stirrups will come directly under the body and the rider can stand in them at will as he would on the floor;
5. Put part of your weight in the stirrup and, pulling your heels down, feel as if the weight in the stirrups actually goes into the heels;
6. Tilt your torso forward from the hips, until your body gets balanced in the stirrups;
7. When leaning forward, keep your torso in a normal, alert position with chest open and head up. Push your buttocks back toward the cantle (no weight in them);
8. Rider's weight should be distributed through hip, knee and ankle. Tension in these springs should increase with increase of the shocks of locomotion, and reach a maximum in jumping . . .

Excerpt taken from *Commonsense Horsemanship* (2nd edition, Arco Publishing, New York 1976) by Captain Vladimir Littauer.

Although the angles of the forward or jumping seat are in sharp contrast to the perpendicular lines of the classical seat, the principle of trying to match the balance of the rider to the balance of the horse is similar. In the extended movements, the horse must be allowed to stretch freely forward; for galloping and jumping he tends to be on the forehand, so naturally he is helped if the rider is leaning forward with the movement, rather than behind it.

The beauty of the forward seat is that it allows the horse total freedom of the back and hindquarters, and this encourages him to remain in balance and develop a steady rhythm – without interference from the rider. The idea here is directed freedom rather than specific control, and the hand should follow the movements made by the horse's head and neck, always maintaining a light elastic contact.

In order to jump more comfortably in a saddle designed to accommodate a shortened stirrup leather with greater angles of the rider's leg, a good jumping saddle or general purpose saddle is obviously preferable to a straighter cut saddle. Most jumping saddles have a knee roll to give added support, although this is not essential. When not in the jumping position, it is important that the rider's weight is spread as evenly as possible over the whole bearing surface; particular attention should be paid to the width of the central channel, to ensure that an adequate passage of air can pass over the horse's back. Severe damage can be done to the horse's back by a saddle which pinches in front or behind, and it is always wise to take professional advice over the fitting of any saddle.

The General Purpose Seat
This term generally denotes a seat which has all the basic ingredients of the classical seat, but the rider will use a shorter stirrup, so that when necessary he can adopt a more forward position for jumping without having to stop and readjust his stirrup leathers. One of the criticisms voiced by instructors concerning the general purpose seat is that it is neither one thing or the other, and can lead to sloppy riding.

This problem should not arise provided that riders remember the importance of keeping their weight central and forward, and never allow it to be directed in pressure points through the back of the buttocks in an armchair position, with the feet thrust forward instead of being directed back and under the rider's weight.

The common feature of all the three seats is therefore:

1. to respect the horse's back
2. to be safe, secure and in control
3. to be in a position of balance which complements the task to which we have put our horse, *ie* to be at one with the horse.

TRAINING THE YOUNG HORSE

It is an unfortunate fact of life that many people with little experience of either the horse's mentality and anatomy, or the importance of a quiet independent seat and the giving of aids, undertake to break and make young horses.

Purposely therefore, I have left all mention of this subject until the end, and it must be recognised that for such an extensive and important topic, much careful preparation and study is required which is beyond the scope of this section. There is, however, a wealth of books which deal exhaustively with the various aspects, and hopefully the guidelines below will give an idea of just how much is involved: it cannot be stressed too strongly that each stage must be taken seriously, allowing plenty of time and effort before passing onto the next.

When training a youngster for the first time, the owner should never be too proud to take advice. It is all too easy to make mistakes, and two heads are generally better than one, especially if there is someone of experience readily to hand.

Foalhood

Few people realise that the training process starts on day one. From careful, fair but firm handling, the young foal gains trust and confidence. From the start, however, he must also learn what is acceptable and what is not, and whilst plenty of handling, handgrooming etc, is good, it is a mistake to indulge in horseplay with the youngster as this only encourages nips and nonsense which are less funny as he gets bigger and stronger.

The first real lesson will be the introduction to the halter. This is best done within the first week, so that the foal becomes used to being led beside his mother. Thereafter it is wise to practise putting the halter on and off regularly until the foal happily accepts the whole process. It is essential that the halter is well-fitting and durable since, later, when you want to tie him to the wall for grooming or picking up feet, he will initially resist. Only if the halter is strong will he discover that there is nothing to be gained by fighting it, and that it is his own actions which bring about punishment.

When the foal learns something new, be quick to reward him with voice and hand; however, titbits are a mistake at this age, and better reserved for the adult horse.

The Three-Year-Old

By the time he is three, if handling in foalhood has been consistent and correct, the young horse should be quiet to handle, groom, lead, tie up and come from field to stable and stable to field. Many people introduce the mouthing bit whilst the young horse is tied up for grooming or enjoying some other pleasurable experience, so the transition to the snaffle bridle will seem natural and normal, particularly if done in the way described below, *ie* at an earlier stage than lungeing and backing.

People have varying ideas as to when to start lunge and ridden work, but it is far better left until after the horse has turned three rather than before. It is quite a culture shock for the young horse – who has spent most of his life turned out in a field – to be brought into the yard and suddenly deprived of his freedom, even for a morning. With the young, unbroken horse therefore, it is a good idea to allow him loose in a safe indoor or outdoor manège for up to 30 minutes per day, before you attempt to think of lunge or ridden work. For safety it is a good idea to put on boots, since it is all too easy for him to knock himself as he lets off steam at the onset of exercise.

After a few days of getting used to this new experience, he can be introduced to the snaffle bridle, and a saddle with the stirrups removed. Use a thick, straight vulcanite or rubber snaffle, taking good care as to the fit and comfort of bit and bridle. It is a good idea to do this after his exercise in the school, when he is less likely to resist. For his introduction to the saddle, use an old one which can be placed lightly on his back, without the girth initially, whilst he is patted and reassured. After a couple of days of this the girth can be done up, but not too tightly, and the horse led around the school or paddock. This should be done in a matter of fact manner with the reins over the head in the normal way. Walk and halt, halt and walk should be taught with short voice commands; everything must remain very calm, and the advantages of working indoors are at this stage considerable. If he has been led successfully as a foal, the young horse should soon get used to being led around the school with his new equipment, and later he can be led in the yard and in and out of his box in the same way.

The First Lunge Lessons

By now the horse is ready to be introduced to the lunge. The purposes of lunge work are many and varied:

1. It introduces the horse to working with his trainer in a controlled way. It accustoms him to his voice, his commands and to general discipline;

RIDING AND SCHOOLING

Foalhood
The three-year-old
The first lunge lessons

A young horse correctly saddled and booted stands quietly waiting for his first hack out. Note the rein is safely secured by the throatlatch, whilst the rope attached to the headcollar is tied in a safety knot.

A well-fitting lungeing cavesson. The browband (which many types do not have) helps keep it in position and the jowl straps keep the cheekpieces from rubbing the horse's eye. The noseband fits snugly but not tightly and is comfortably padded. If it is too loose it will slip round, pulling the whole cavesson with it, lessening control and making the horse uncomfortable.

2. It encourages free forward movement, and promotes correct muscular development and the flexion and easing of the joints;

3. It supples the horse laterally on both sides by stretching and shortening the muscles of the neck, back and abdomen: on the right rein, the left muscles are stretched and the right shortened or contracted; on the left rein, it is the reverse process. This occurs because the horse is bent on the circle.

4. Finally, as long as you proceed with patience and tact, it establishes in the horse's mind the superiority and consistency of his master or mistress. This is very important to achieve before ridden work is undertaken.

So far the equipment has been headcollar, boots, and a well-fitting saddle and bridle; now it is time to add the following:

lunge cavesson; lunge roller; lunge line; lunge whip; side reins

Some people prefer to use just the cavesson with a snaffle bit attached; others, particularly on the continent, keep the bridle on (with the reins removed, or at least safely secured out of the way) and then place the cavesson on top. The lunge-line is attached to the middle ring of the cavesson noseband which should be well padded and close-fitting.

If the horse has already been introduced to the saddle (as described) it is a good idea to continue his education with the saddle, and just place the roller on top. Others, however, prefer to start lungeing with only the roller.

The first lunge lesson is preferable without the whip, and with the horse led on the outside by someone whom he knows and trusts. Gradually, the helper relinquishes the horse to the trainer, so that by the third lesson he is working independently. If at any time, however, the horse seems reluctant to remain on the circle, the helper should help to re-establish the pattern. At this stage it is most important to be content with very little at a time. The first and second lessons are best conducted in walk only. The trainer can bring the whip into the circle by the third lesson, picking it up and putting it down so that the horse becomes accustomed to it, before actively using it to guide the horse forward.

From this point in training, everyone has his own way of proceeding; some may therefore disagree, but I feel it is often safer to apply the side-reins before asking the horse to trot on the lunge. These connect the rings of the snaffle to the roller, but must be long enough for the horse to walk normally with head outstretched so that his freedom is not impaired. Their purpose is to replace the rider's hand and allow the horse a feeling of contact. The inside side-rein should be one to two holes shorter than the outside, depending on the size of the circle.

The working trot is the main gait of the lunge lesson. Introduced slowly and with plenty of walks in between and assistance from the helper, the horse should not use this increase in gait as an opportunity to fly wildly round, as is too often seen. By this time the trainer should have accustomed the horse to the whip, and the idea is to keep the horse on a triangle of control (see diagram) between the lunge-line – which should never be allowed to slacken – and the whip.

In this way the horse grows used to the checking aid of the hand and the driving aid of the whip (which replaces the rider's legs). As soon as the trot is calm, forward and unhurried, the canter may be introduced. As the horse grows more supple and confident, the side-reins may be slightly shortened as necessary.

Successful lungeing makes the whole process of backing the young horse very much easier.

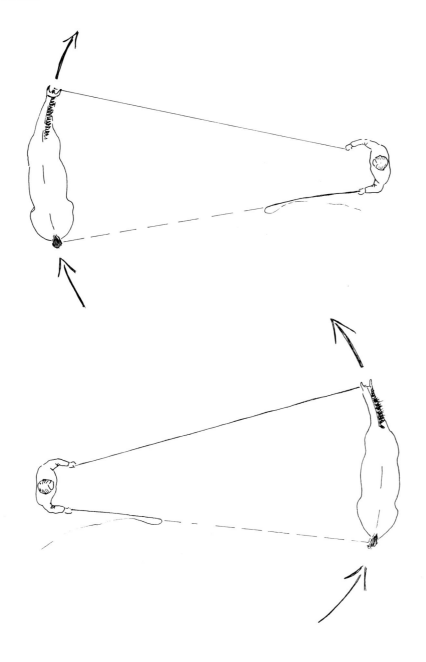

Once the horse is confident on the lunge he must be maintained through a triangle of control formed by the lunge-line and the whip, with the trainer always pushing the horse ahead of the line of his shoulder.

The horse must be worked equally to right and left so that he becomes laterally supple on both sides.

Backing

It is essential that the horse trusts both the trainer and helper before he is backed. Again, it is best to borrow an indoor manège for the first proper lesson – the expense will pay dividends. After the usual lunge lesson, when the horse has finished quietly in walk and been rewarded, the roller is removed, but the lunge cavesson maintained. The reins should now be untied and should lie on the horse's neck in the normal riding position; the helper should stand at the horse's head, securing him lightly with both rein and lunge-line. The trainer should stroke and talk to the horse in the normal way; when everything seems calm, he can have a leg-up from the helper, or if possible a third person, so that he is lying across the saddle. This is done several times until the horse accepts the new weight in the saddle.

Once the horse accepts this calmly, the rider can try dropping quietly into the saddle (which with very nervous horses may take a few days). As soon as he is settled the helper should walk the horse forward onto the usual lunge circle, and the rider should lean forward a little, giving no aids and taking great care to keep his weight well forward.

Often the horse will start trotting. This is not yet desirable, but it is important the rider gives no sign of unease by going against the movement and he should therefore maintain his forward balance and rise gently to the trot. After a couple of circuits, the helper should encourage the horse to return to walk. If all this goes smoothly and calmly, as it should, the rider should then slip off and make much of the horse with reassuring words and pats. This will be quite adequate for the first day.

The next day will be similar, but the rider may practise getting on and off two or three times. His role is still passive, and he must always make his movements quiet and light, taking care not to put undue pressure on the horse's back; everything should be made as natural as possible for the horse. By the third or fourth lesson the lunge-line can be taken off, and once the horse is happy on the usual circuit, he can be taken quietly to other parts of the school. Now the rider is in control, but it is a good idea if a sedate, older horse can be present in the early days of ridden work. If he sees another horse working calmly round the manège, it will give the youngster confidence and make it easier for him to accept all the strange things which are happening.

The young, newly-backed horse must be encouraged to step out on a long, gentle rein. With the rider's seat lightly forward in the first few sessions, the horse will gradually be able to accept a more upright position and the introduction of subtle aids which will help him to come off the forehand and accept more contact in the rein.

Mounted work should be interspersed as much as possible with continued lunge work (without a rider), and if possible some loose schooling or long-reining, both of which are an art in themselves. The young horse will not build up firm muscles if he is ridden for too long too soon; hence the value of other means of exercise particularly if there is no field or paddock available for turning him out after work.

Loose Schooling

Loose schooling should be done with a helper. The equipment is similar for lungeing, though both the trainer and helper should carry a lunge-whip. The horse is booted or bandaged as for lungeing, wears a snaffle bridle with the reins rolled up and secured under the throatlash, a cavesson and a roller. It is helpful to place four cones in each corner of the school about 8ft away from the track. The horse is then encouraged outwards by the people on foot; soon he will respect the cones and want to work to the wall. He should be allowed complete freedom of movement until he gets used to staying out on the track.

Eventually the horse will settle, and then it will be time to put on the side-reins. These pass from the rings of the snaffle to the roller, long enough for the horse to stretch his neck, but not so slack that there is no contact.

With sufficient practice, voice control and reward, horses will eventually work happily in this way in the gait and speed required by the trainer. The benefit of the side-reins is to allow the horse to stretch forward and down, thus rounding and strengthening the back. At a much later stage cavaletti may be introduced, and this not only helps with future jumping, but also etablishes rhythm and balance. A word of warning, however: none of this work should be undertaken without due preparation, and before introducing poles or jumps of *any* description, further reading on the subject is essential. Every horse is different, and size and length of stride plays a vital part in how you lay out your school.

Every lesson – whether in loose schooling, lungeing or ridden work – should end by taking off the side-reins (if used) and quietly walking the horse in hand round the school. This re-establishes his contact with his trainer, and gives him time to relax and dry off.

Lunge and loose school work should last initially for about 15 minutes. Gradually these sessions can be extended from 20 to 25 minutes, but it is vital always to take plenty of time with each new exercise, to end quietly, and always on a good note. It is at this stage that reward with hand, voice and the odd sugar lump begins to play a major part, so that the horse always associates hard active work with appreciation and reward. This is the vital ingredient which encourages the horse always to give of his best in his future relationship with his rider; and on the rider's part, nothing should be taken for granted.

Long-Reining

This is an invaluable method of working the horse in the school, and later, introducing him to the outside world before riding him out. It is important to have a helper until the horse becomes used to this work, but if well done the trainer will eventually have infinitely more control over his horse than he does on the lunge. Equipment is similar to lungeing except that there are two lunge lines, each with a buckle at one end and preferably a knot at the other to prevent it from slipping from the trainer's grasp. The buckled ends pass from the rings of the snaffle (some people start them on the noseband and then through the rings of the bit) through the rings of the roller or through the stirrup irons (if you are using a saddle) which are initially adjusted and secured to a height about 8in higher than the horse's elbow, and finally into the trainer's hand. The trainer works his horse from a position equivalent to a good horse's length back from the horse's tail.

If the horse has already been lunged and loose schooled successfully, the transition to the long-reins should not be too traumatic, although some horses

This horse is nicely established in long-reining, and shows an attentive response to his handler's light use of the rein.

object at first to the feeling of the rein around the quarters – in this case it may be necessary to pass the outside rein over the horse's neck just behind the withers, until he is happier to feel the contact round the quarters.

Long-reining is most definitely an art, and should not be attempted without further reading and ideally, help from an expert. The skill lies in the handling of the reins and the whip; when competently done, a horse can be made to perform all the normal ridden exercises, from basic gaits to advanced dressage airs, displaying great freedom and beauty.

Forward and Straight!

By now the young horse will be ready to commence his full education. This should be divided into four phases:

1. Preliminary lunge work* (to let off excess steam and energy and to supple him on the circle);

2. Ridden walk on a long, loose rein (to stretch, relax and supple the muscles and to settle the horse mentally);

3. More active ridden work (to establish the aids, the basic gaits, to build muscles and to create impulsion behind the saddle);

4. Dismounted, walk in hand (to reward, re-establish personal contact, relax, and possibly dry off).

Most horses need 15 minutes on the lunge if this is to achieve the objectives, although a phlegmatic horse will require less than an energetic, highly strung one.

The second phase is most important for all horses. It is the loosening, suppling phase, although the rider must not use this as a relaxation period for himself. The purpose is to teach the horse to step out and stretch himself freely forward with an active rounded back. To start with, at least 10 minutes is advisable each day; after the second month, this can be reduced to about 5, but preferably never less than this. It is a discipline which can reap great benefit if carried on into the second year of training and beyond.

*Some will prefer to replace this stage with loose schooling or long-reining.

For the first few days the third phase, the work proper, can be omitted as the horse is still getting used to his rider, and the aids are still foreign to him. Gradually however, the more active work is introduced, 5 minutes to start with, later 10, and by the end of the third month, the healthy young horse may manage 15 minutes – although this must still be liberally interspersed with short rest periods on a loose rein. The programme by now may look like this:

15 minutes on the lunge;
10 minutes ridden on a long, loose rein;
15 minutes of the work proper (with short rest periods);
5 to 10 minutes relaxation, walk in hand to dry off.

The horse is naturally very unbalanced in the early stages of ridden work. It is important to keep him moving freely forward and straight at all times to combat his natural crookedness, and large half circles should be used to change the rein. Many people make the mistake of trotting fast to push the horse on, but this often results in the quarters straying in because the rider cannot control them properly, and the horse leans too much on the forehand. The latter often leads to a build-up of tension in the neck and back which may cause a buck and the rider falling. Nothing is more unsettling for the young horse, so it is most important that falls happen as seldom as possible, and preferably never at this formative stage. If the rider feels overfaced, another more experienced person should take over. I have seen horses, particularly the clever ones, which have never completely recovered their manners, having learned too early in life how to rid themselves of an unwanted burden.

Until the horse develops the muscles of the neck, back and quarters, a process which will take several months, the rider must try to sit as lightly and as quietly as possible. He should rise gently to the trot, keeping his body inclined forward of the vertical and his hands low enough to form a straight unbroken line from the elbow

BELOW LEFT:
Most horses are naturally crooked in all gaits and especially at the canter.

CENTRE:
Riding a horse correctly through corners and on circles will help combat this problem. Note this rider has made it easier for his horse to bend correctly by allowing his shoulders to turn very slightly to complement the horse's shoulders, i.e. outside shoulder comes a little forward. He will also increase the downward pressure into his inside stirrup so that he goes *with* the direction of the horse and not against it.

RIGHT:
Well ridden corners and large circles encourage the young horse to achieve straightness in all gaits.

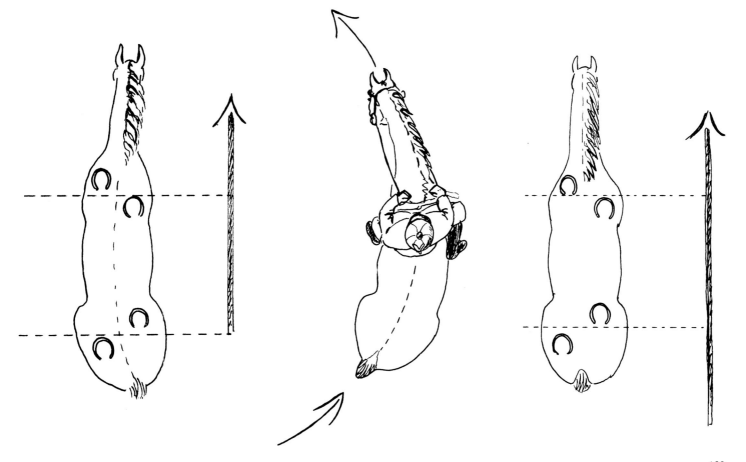

This young mare has taken this small obstacle willingly in a free-going manner. Christa, her rider, has folded down and gone with the mare's movement and although she has just lost contact with the bit it is far better to do this slightly than to stab the horse in the mouth. It is, in fact, very difficult to maintain a slight contact with the bit at *all* times, as can be seen by watching any show jumping class at any level.

through the reins to the horse's mouth. It is a severe mistake to hold the hands too low or to attempt to fix the horse's head down. Roundness and swing will only come with increased activity behind the saddle, and not from an artificial contact with the hands.

Once the young horse is settled in the school he may be gradually introduced to riding out. A quiet companion is the best means of ensuring an uneventful first ride out, and riders should be content to remain in walk for the first few hacks into the countryside. Gradually, short spells of slow trotting may be introduced, but cantering should still be avoided until the horse has learned to respond more fully to the aids and is used to his surroundings. To ask him to canter in a strange environment will probably excite him greatly, but this may result in his rider pulling and tugging at his mouth which is most unpleasant for him. It is at this stage that the young horse may learn to throw his head up and avoid the bit, which can later lead to many problems.

THE CLASSICAL WAY FORWARD

The work proper, whether in the school or out, requires much thought and planning from the rider. Simple exercises can be devised (similar to those found at novice dressage stage) as well as gait transitions, but the emphasis must always be on forward impulsion, straightness and a calm, steady approach. It should be quite clear to the horse when he is required to go loosely forward in a long, low outline, and when he is expected to pick himself up for the more active work. For this, the rider will sit up a little more and make a positive contribution through his own balanced posture as well as light leg and stick aids, to help the horse readjust his balance. And remember that it is of little use encouraging the horse into his stride through increased energy behind, if we always allow it all to escape through the forehand: if 'long and low' is pursued for too long, the results can be deleterious.

Whole books have been written about the question of rebalancing the horse so as to allow the weight to come a little further back; this in turn allows the hindlegs to generate more power and energy, which is then channelled through an unconstricted back into a light even contact. The whole process certainly cannot be described in a few paragraphs, and riders must be prepared to tackle this stage of training with patience and understanding. Only if this is correctly done, will the horse learn to work in self-balance and eventually develop good collection and extension.

Riders should be guided by the works of recognised masters rather than be led astray by the latest perpetrator of 'natural methods'. The classical school has been using and improving upon natural methods since the time of Xenophon, and there are many excellent books on the market. Remember that the truly natural method is the one which enables the horse to rebalance himself with the rider's weight on top. *Only then can he really move as Nature intended,* ie with spring, elevation, pride and impulsion. This must never be forgotten. There is nothing natural at all about horses which progress round and round a school with all the weight on their forehand, and their heads fixed somewhere between their knees.

Every horse is different, and programmes will vary depending upon his age, conformation, and the time the rider is able to give him. The main thing is to take one stage at a time, to concentrate on making things easy for the horse, to be consistent and fair, and always to reward when the horse has tried or achieved.

A final thought comes to us from Commandant Jean Saint-Fort Paillard, a brilliant and most understanding horseman who wrote a valuable book called *Understanding Equitation,* suitable for all thinking riders, irrespective of which discipline they wish to follow with their horses.

Taking into account the gifts of each individual horse, the degree of intelligence it acquires will correspond to man's ability to help it attain it. In successful cases, this can place the horse in a very honourable position among the animals generally considered to be intelligent.

Next time you hear someone calling his horse 'stupid!' 'obstinate!' 'unco-operative!' – think on this.

There is every reason to reward your horse when he has worked well. Personally, I always have a packet of Polos in my pocket during every schooling session. Wisely given, rewards build up trust and confidence. This is my Lusitano-Arab stallion, Palomo Linares, a schoolmaster in High School work. (Photo: Madeleine McCurley)

Happiness is a horse which understands every signal, every minute request made by its rider, which can be performed without pain, stress or discomfort. Horses love to please. Here, the author of this part of the book rides her Lusitano mare, Andorinha, in a forwardly active, engaged but clearly relaxed trot. (Photo: Madeleine McCurley)

CONCLUSION

As stated in the Preface, this book is meant as an introduction for the novice rider to the skills of horse care and management, and there is also a very valuable section on riding and schooling. Certain topics have been included which are not traditionally the field of the novice – breeding and schooling horses – as it is felt that a basic knowledge of these subjects can only help to increase the reader's understanding of horses in general, even though he or she may not yet be directly involved in these areas.

As your interest deepens and you read other books, take practical instruction in both care and equitation and perhaps undertake formal courses of study, maybe with examinations or a career in view, you will realise that experts often disagree, and you will certainly form your own opinions based on your increased experience.

I firmly believe that the way to success is to always try to look at things from your horse's point of view. The horse is a magnificent animal who often attracts to himself (or herself) people who only want to be associated with horses because of the kudos and prestige the connection brings; because of the feeling of power conveyed by riding and controlling (we hope) an impressive animal much stronger and faster than any human; or because of the supposed snob value which horse riding and ownership still seem to imply. If the horse or pony is the type who wins rosettes, this, too, often attracts the wrong sort of people – those who only own horses as a means to compete and beat others. Competition is fine provided it is kept in perspective. Because there is so much emphasis on competition in today's horse world, there are people who only look after horses 'properly' so that they can be maintained in a suitable condition to compete and win, with little regard for the actual well-being of the horse as an individual personality.

As this book is specially for novices in the horse world, I very much hope it will encourage them to start off and continue regarding the horse himself as more important than anything he may do for us; not as a means to an end, but as the end purpose itself – a trusting partnership between horse and human – with whatever activities that partnership may enjoy as secondary bonuses. Good, caring horse management and riding will greatly enhance both factors.

FURTHER READING

Near the front of this book is a list of my own books and also, elsewhere, a list of equestrian books published by David & Charles, and I hope on both these lists readers will find other titles to dip into or add to their bookshelves. A logical follow-on to *Horse Care and Riding: a thinking approach* is my book *Effective Horse and Pony Management: a failsafe system*, also published by David & Charles. This list below consists of other author's books which I personally recommend (even though I may not agree with every little point in them!) and which I believe readers will benefit greatly from reading. First, two reference books:

Rossdale, Peter D. and Wreford, Susan M. *The Horse's Health From A to Z: an equine veterinary dictionary* (David & Charles, 1989) ISBN 0–7153–9266–2. A comprehensive veterinary dictionary for the horse enthusiast, giving a good general introduction to and explanation of the many technical and scientific terms which can often baffle the newcomer, not to mention the expert.

Summerhays, R. S. *Summerhays' Encyclopaedia for Horsemen* (Threshold, 1988) ISBN 0–901366–44–7. A comprehensive reference work giving general information in alphabetical order form on just about any topic in the horse world. Invaluable to novice and expert alike – especially if you are afraid to ask 'silly questions'!

The following are instructional books with good indexes so they can be used for reference as well as for reading straight through, as required:

Hartley Edwards, E. *Saddlery* (J. A. Allen, 1990) ISBN 0–85131–151–2. A revised edition of the author's 1963 classic. Really makes you think about what you expect your horse to put up with! Most informative.

Kiley-Worthington, Marthe. *The Behaviour of Horses in relation to Management and Training* (J. A. Allen, 1987) ISBN 0–85131–397–3. A really super book which drills home many things which have needed saying for years. The author is a zoologist and ethologist specialising in equine behaviour and communication.

She is also a practical horsemaster, breeder and competitor. I feel this book should be read, and its principles applied, by everyone who has anything to do with horses, particularly examining organisations, their instructors and examiners.

Loch, Sylvia. *The Classical Seat* (D. J. Murphy (Publishers) Ltd, revised edition 1991 (ISBN 0–04–440177–9) Published by *Horse and Rider* magazine, this small but very concise book expands on the chapter the author has written for *Horse Care and Riding*. I recommend it absolutely to those wishing to develop a harmonious riding technique horses.

Loch, Sylvia. *Dressage: The Art of Classical Riding* (Sportsman's Press, 1989) ISBN 0–948253–46–0. A superb and fascinating history of classical equitation from the ancient Greeks to the present day. Instructional both technically and historically, with many fascinating anecdotes about past and present masters of the art. Above all, it is encouraging and shows that with application and average talent anyone can aspire to some level of classical equitation.

McCarthy, Gillian. *Pasture Management for Horses and Ponies* (Blackwell Scientific Publications, 1987) ISBN 0–00–383330–5. The only really comprehensive and up-to-date book on the subject. Grassland is an important and wasted resource, neglect of which costs the horse industry many thousands of pounds per year. Definitely a book to take note of.

Pavord, Tony and Fisher, Rod. *The Equine Veterinary Manual* (Crowood Press, 1987) ISBN 0–946284–29–6. An excellent book explaining the physical workings of the horse, what can go wrong with them and what can be done about it. Up-to-date, reliable and highly recommended.

Pilliner, Sarah. *Getting Horses Fit* (Blackwell Scientific Publications, 1986) ISBN 0–00–383197–3. Getting and keeping horses and ponies fit and sound for their work, however moderate that might be, is essential to good horsemastership and this book explains everything in considerable detail yet in a form which is fairly easy to understand even if you weren't any good at biology at school!

Rees, Lucy. *The Horse's Mind* (Stanley Paul, 1984) ISBN 0–09–153660–X. This book is another in the mould of Dr Marthe Kiley-Worthington's book recommended above but covers the subject from a slightly different angle. Again, the same comments apply.

Vogel, Colin J. *The Stable Veterinary Handbook* (David & Charles, 1990) ISBN 0–7153–9218–2. Complimentary to the *Equine Veterinary Manual* recommended above, this is more of a first aid-cum-nursing book telling owners what they can and should do for their horses themselves, with or without veterinary supervision, as the case may be. Up-to-date and reliable, unlike some vet books currently on examining bodies' reading lists!

Wanless, Mary. *Ride With Your Mind* (Methuen, 1987) ISBN 0–413–40820–5. A book to read, absorb and apply before you get into too many bad habits, and even if you have already done so. This book will guide you towards a truly sensitive partnership with any horse, one in which you are going with your horse in every way instead of resisting him and forcing him to resist you. These methods have helped many, many riders and will surely help you.

Finally, two books which you may prefer to read when you have gained more experience and have studied those already listed. They are, in my opinion, not suitable for real novice riders but should be studied once you have a sound, basic grasp of horsemanship.

Smythe, R. H. *The Horse: Structure and Movement*, second edition revised by P. C. Goody (J. A. Allen, 1972). Don't be put off by the blocks of text, the Latin names and the drawings of bones and muscles. The knowledge to be gained from this book is substantial and you'll view your hours in the saddle in a different light after reading this. It *is* perfectly understandable – but don't start reading it when you're tired!

Tellington-Jones, Linda and Bruns, Ursula. *An Introduction to The Tellington-Jones Equine Awareness Method: The T.E.A.M. Approach to Problem-Free Training*. Another excellent book getting away from the stereotyped Establishment view of handling and training horses. A comparatively new method which takes great account of the psychology of the horse, the T.E.A.M. method is spreading in popularity around the world and coming up with great results. You're bound to get a lot out of this book.

This is a fairly short but carefully selected, comprehensive reading list which will stand you in good stead for the time being. A longer list is given at the end of my book *Effective Horse and Pony Management*. There are thousands of horse books on the market and the choice is bewildering. It can be especially confusing to novice riders to read one thing in one book and an opposing view in another written by an equally eminent expert, and to find that one's instructor disagrees with something you have read and which you'd like to try. I feel that you should read and consider all you can and remember that horses are very much individuals – what works for one may not work well for another.

Above all, *remain open-minded* and be ready and willing to adapt your views and practices in the light of your increasing knowledge. Bigotry, self-righteousness and sheer block-headedness are serious enemies of progress and of a harmonious relationship with the horse.

INDEX

ACKNOWLEDGEMENTS

A large part of this book has been written by three people whose specialised knowledge brings an authority these particular subjects might otherwise have lacked. I am extremely grateful to them for taking the time to enhance the book in this way, and know that their contributions will prove of significant and practical help to readers at all levels of knowledge.

Firstly, there is Gillian McCarthy, BSc (Hons), an internationally respected equine nutritionist and management consultant who undertook to write the substantial feeding section in Part 4, in spite of a long-term and very debilitating illness. Then Sylvia Loch, an international judge who lectures on, demonstrates, and teaches classical riding; in Part 7 she has written an extremely practical and relevant explanation of how to ride in true harmony with your horse – even if necessarily this has had to be somewhat 'potted'. And finally Simon Wolfensohn, MA, VetMB, MRCVS, who took time off from a very busy veterinary practice to write Part 2 and the *Veterinary Aspects* section of Part 6, to bring to the book the latest in veterinary thinking and practice.

We have all tried to give the most up-to-date facts and thinking on the various aspects of management most necessary for novices to understand if they are to progress to a higher standard. Many old principles are scotched and replaced by more appropriate ones, and the latest thinking and the reasoning behind it is presented in what I hope is a clear fashion, easily understood.

Gillian McCarthy would herself like to thank her sister Clare, and also Joan Williams, for their help in the preparation of her manuscript – and also the author, but why I don't know!

My thanks are also due yet again to the artist, Joy Claxton, who has illustrated several of my books with seemingly interminable patience, expertise in both art and matters of equestrian knowledge, and diligence. This one is no exception, and what I have failed to 'catch' with my camera she has made up for in pen and ink in her usual individual and attractive fashion.

And now, it seems quite inadequate to merely thank my editor, Sue Hall, the person who has held this project together, for her truly incredible patience and tact when dealing with a book which has had more than its fair share of not just hiccups but impacted colic. Whatever they pay her she's worth twice as much.

Another job which requires lots of patience and co-operation is staging and setting up photographs, and for this often trying and boring job I wish to thank most sincerely Mrs Frances Phillips and friends of Holford Hall, Plumley in Cheshire, Mrs Margaret Mitchell and Miss Antonia Mitchell of Swordhill Training Centre, Stalmine, near Blackpool, Lancashire and Miss Nicola Smith of Renards Livery Stables, also in Stalmine; they all helped most generously with time and facilities, which meant I could produce the majority of the photographs for this book comparatively quickly and with the minimum of fuss.

I would also like to thank Barrie Hosie for his very patient help and advice regarding photographs, and equestrian photographer Bob Langrish for making up my deficiencies.